Communication Technology and Human Development

Communication Technology and Human Development

Recent Experiences in the Indian Social Sector

Avik Ghosh

SAGE Publications
New Delhi • Thousand Oaks • London

Copyright © Avik Ghosh, 2006

All rights reserved. No part of this book may be reproduced or utilised in any form or by any means, electronic or mechanical, including photocopying, recording or by any information storage or retrieval system, without permission in writing from the publisher.

First published in 2006 by

Sage Publications India Pvt Ltd
B-42, Panchsheel Enclave
New Delhi 110 017
www.indiasage.com

Sage Publications Inc
2455 Teller Road
Thousand Oaks, California 91320

Sage Publications Ltd
1 Oliver's Yard, 55 City Road
London EC1Y 1SP

Published by Tejeshwar Singh for Sage Publications India Pvt Ltd, Photo-typeset in 10/12 Life BT by Siva Math Setters, Chennai and printed at Chaman Enterprises, New Delhi.

Library of Congress Cataloging-in-Publication Data

Ghosh, Avik.
 Communication technology and human development : recent experiences in the Indian social sector / Avik Ghosh.
 p. cm.
 Includes bibliographical references and index.
 1. Communication—India—Social aspects. 2. Communication and technology—India. 3. Information technology—Social aspects—India. I. Title.

HN690.Z9I564   303.48'33095409045—dc22   2006   2005033243

ISBN: 0-7619-3438-3 (Hb)   81-7829-584-9 (India-Hb)
       0-7619-3439-1 (Pb)   81-7829-585-7 (India-Pb)

Sage Production Team: Sarita Vellani, Shinjini Chatterjee, Sanjeev Sharma
and Santosh Rawat

For my parents

Contents

List of Tables, Figures and Boxes	8
List of Abbreviations	10
Acknowledgements	15
Introduction	17

Section I: Historical Perspective

1. Communication and its Role in Development	25
2. Search for an Alternative Development Paradigm	40
3. Alternative to Broadcasting: Some International Experiences	49
4. Networks and Movements	59
5. Communication and Human Development: A New Approach	69

Section II: Recent Experiences in India

6. Use of Communication for Literacy and Empowerment	81
7. Population: Bringing about Behaviour Change	147
8. Reaching Development to the Rural Poor	201
Conclusion: Communication Challenges in India	255
Bibliography	276
Index	279
About the Author	291

List of Tables, Figures and Boxes

TABLES

6.1	The Search for Volunteers	119
7.1	Organisational Aspects	153
7.2	Demographic Issues	155
7.3	Matrix of RCH Behaviour Change Objectives	176
8.1	Training Activities	209

FIGURES

5.1	Communication and Development Model	72
5.2	Behaviour Change Communication Process	75
5.3	Basic Flow-Diagram of the BASNEF Model	76
5.4	Communication Planning (P-Process)	77
6.1	Page from *Khilti Kaliyan* Primer (Lesson on *Bari Ladki*)	95
6.2	Akashvani Pathmala	105
6.3	The First Lesson of PREAL	109
6.4	Chauraha TV Lessons	110
6.5	Copy of Hindi Press Advertisement for Literacy	120
6.6	A Tribal Couple Watching a *Kalajatha* Performance for Literacy	134
6.7	Young Performers at a *Kalajatha*	135
6.8	Village Procession and Rally during the Bharat Gyan Vigyan Jatha	135

6.9	Broadsheets in Different Languages for Neoliterate Persons	142
6.10	Women Learners Reading Lessons	143
7.1	Poster Promoting NSV for Men	181
7.2	Attempt to Evolve the 'Red Triangle'	182
8.1	ANSSIRD SATCOM System Diagram	206
8.2	A GP Member at a Field Centre	211
8.3	Sanitary Latrines and Squatting Platforms	227

BOXES

1.1	The Information Model	28
1.2	The Emergency—Loss of Credibility	37
6.1	*Padhna Likhna Seekho*—Literacy Song by Safdar Hashmi	101
6.2	*Chauraha*—An Instructional TV Serial	111
6.3	The NEEM/UEE Campaign	123
7.1	Orissa—An Exception	187
8.1	Films Used for Training	212
8.2	Rural Sanitary Marts (RSMs): The Medinipur Story	228
8.3	From the Field Notes of Malini Ghose	239
8.4	From the Field Notes of K.K. Krishna Kumar	242

List of Abbreviations

AE	Adult Education
AEC	Adult Education Centre
AIR	All India Radio
ANC	Antenatal Care
ANM	Auxiliary Nurse/Midwife
AWW	Anganwadi Workers
ANSSIRD	Abdul Nazir Sab State Institute for Rural Development
ARWSP	Accelerated Rural Water Supply Programme
BASNEF	Behaviour, Attitudes, Social Norms and Enabling Factors
BCC	Behaviour Change Communication
BDO	Block Development Officer
BEE	Block Extension Educator
BGVJ	Bharat Gyan Vigyan Jatha
BGVS	Bharat Gyan Vigyan Samiti
BPL	Below Poverty Line
CBC	Communication for Behaviour Change
CBO	Community-based Organisation
CEERI	Central Electronic Engineering Research Institute
CENCIRA	(tr.) National Centre for Land Reform Education and Research
CENDIT	Centre for Development of Instructional Technology
CESPAC	(tr.) Audiovisual Teaching Services Training Centre
CIET	Central Institute of Educational Technology
CIG	Common Interest Group
CMO	Chief Medical Officer
CMS	Centre for Media Studies
CNA	Community Needs Assessment
CRSP	Central Rural Sanitation Programme
CSE	Centre for Science and Environment
CSO	Civil Society Organisation

LIST OF ABBREVIATIONS

CSSM	Child Survival and Safe Motherhood
CWC	Concern for Working Children
DAE	Directorate of Adult Education
DAEO	District Adult Education Officer
DDK	Delhi Doordarshan Kendra
DECU	Development Education and Communication Unit
DFW	Department of Family Welfare
DH&FW	Department of Health and Family Welfare
DHO	District Health Officer
DLC	District Literacy Committee
DMEIO	District Mass Education and Information Officer
DPEP	District Primary Education Programme
DPIP	District Poverty Initiative Programme
DRS	Direct Reception Set
DW&CD	Department of Women and Child Development
DWCRA	Development of Women and Children in Rural Areas
EAG	Empowered Action Group
EC	European Commission
ECCE	Early Childhood Care and Education
EDLC	Ernakulam District Literacy Committee
EFA	Education For All
EPI	Expanded Programme of Immunisation
EWE	Education for Women's Equality
FICCI	Federation of Indian Chambers of Commerce and Industry
FP	Family Planning
FWP	Family Welfare Programme
GNP	Gross National Product
GOI	Government of India
GOWB	Government of West Bengal
GP	Gram Panchayat
GRAMSAT	Satellite for Rural Communication
HDR (MP)	Human Development Report (Madhya Pradesh)
HLL	Hindustan Latex Limited
HRD	Human Resource Development
HtH	House to House
IAY	Indira Awas Yojana
ICDS	Integrated Child Development Services
ICPD	International Conference on Population and Development

IDS	Institute of Development Studies
IEC	Information, Education and Communication
IGNOU	Indira Gandhi National Open University
IHL	Individual Household Latrine
IIMC	Indian Institute of Mass Communication
ILD	International Literacy Day
ILY	International Literacy Year
INSAT	Indian National Satellite System
IPC	Interpersonal Communication
IPCL	Improved Pace and Content of Learning
IPTA	Indian People's Theatre Association
IRDP	Integrated Rural Development Programme
ISP	Intensive Sanitation Programme
ISRO	Indian Space Research Organisation
ISST	Institute for Social Studies Trust
IUD	Intra-Uterine Device
JFED	Jharkhand Forests and Environment Department
JFM	Joint Forest Management
JHU/PCS	Johns Hopkins University/Population Communication Services
KAP	Knowledge, Attitudes and Practice
KSSP	Kerala Shastra Sahitya Parishad
MCH	Maternal and Child Health
MEM	Mass Education and Media
MHRD	Ministry of Human Resource Development
MIS	Management Information System
MLA	Member of Legislative Assembly
MMR	Maternal Mortality Rate
MNP	Minimum Needs Programme
MEIO	Mass Education and Information Officer
MO	Medical Officer
MOEF	Ministry of Environment and Forests
MOHFW	Ministry of Health and Family Welfare
MPFL	Mass Programme of Functional Literacy
MSS	Mahila Swasthya Sangh
MTR	Mid-Term Review
NBA	Narmada Bachao Andolan
NCC	National Cadet Corps
NCERT	National Council of Educational Research and Training

LIST OF ABBREVIATIONS

NEEM	National Elementary Education Mission
NFDC	National Film Development Corporation
NFE	Non-Formal Education
NFHS	National Family Health Survey
NFWP	National Family Welfare Programme
NGO	Non Governmental Organisation
NIAE	National Institute of Adult Education
NIAHRD	National Institute of Applied Human Research and Development
NID	National Institute of Design
NIHFW	National Institute of Health and Family Welfare
NLM	National Literacy Mission
NLMA	National Literacy Mission Authority
NPE	New Policy on Education
NPP	National Population Policy
NPPF	Non-party Political Forces
NRR	Net Reproductive Rate
NSS	National Social Service
OB	Operation Blackboard
ORG	Operations Research Group
P&RD	Panchayats and Rural Development
PC	Production Centre
PF	Protected Forest
PFT	Process Facilitation Team
PHC	Primary Health Care
PHED	Public Health Engineering Department
POA	Programme of Action
PRA	Participatory Rural Appraisal
PRC	Population Research Centre
PPI	Pulse Polio Immunisation
PREAL	Project in Radio Education for Adult Literacy
PRI	Panchayati Raj Institutions
PSC	Programme Support Communication
RCH	Reproductive and Child Health
RF	Reserve Forests
RH	Reproductive Health
RKMLSP	Ramakrishna Mission Lok Shiksha Parishad
RSM	Rural Sanitary Mart
RTI/STD	Reproductive Tract Infection/Sexually Transmitted Diseases

SAARC	South Asian Association for Regional Cooperation
SAC	Space Applications Centre
SAHMAT	Safdar Hashmi Memorial Trust
SATCOM	Satellite Communication
SC/ST	Scheduled Caste/Scheduled Tribe
SDAE	State Directorate of Adult Education
SHG	Self-help Group
SIHFW	State Institute of Health and Family Welfare
SIPRD	State Institute for Panchayats and Rural Development
SITE	Satellite Instructional Television Experiment
SGSY	Swarnajayanti Gram Swarozgar Yojana
SMCR	Source–Message–Channel–Receiver
SR	Sector Reforms
SRC	State Resource Centre
SWRC	Social Work and Research Centre
TBA	Trained Birth Attendant
TFA	Target-free Approach
TIFAC	Technology Information and Forecasting and Assessment Council
TLC	Total Literacy Campaign
TOR	Terms of Reference
TP	Tehsil/Taluk Parishad
TRYSEM	Training Rural Youth for Self-Employment
TSC	Total Sanitation Campaign
UEE	Universalisation of Elementary Education
UPI	Universal Programme of Immunisation
UNESCO	United Nations Educational, Scientific and Cultural Organisation
UNFPA	United Nations Population Fund
UNICEF	United Nations Children' Fund
UPE	Universalisation of Primary Education
USAID	United States Agency for International Development
VEC	Village Education Committee
VEDC	Village Eco-Development Committee
VFMPC	Village Forest Management and Protection Committee
WCD	Women and Child Development
ZP	Zila Panchayat
ZSS	Zila Saksharta Samiti (District Literacy Committee)

Acknowledgements

I would like to acknowledge the enormous contribution made by my teachers, colleagues and other professional associates towards my evolution as a practitioner in the field of development communications. In particular, I would like to remember Akhila Ghosh, Anil Srivastava and Rajive Jain. Together we founded Centre for Development of Instructional Technology (CENDIT) in 1972 and ideas on the role of communication technology in development in India were shaped through our early endeavours in CENDIT.

Later, when I joined the National Literacy Mission (NLM) as a Media Consultant in the Directorate of Adult Education, New Delhi I benefited from close association with Ashoke Chatterjee and Gerson da Cunha. Their experience in private sector marketing and communications with its emphasis on research and accountability helped me in planning communication strategies for development programmes.

Various international agencies like the World Bank, United Nations Children's Fund (UNICEF), United Nations Population Fund (UNFPA), etc., as well as several Central and state government departments provided me with opportunities to work on developing communication strategies, preparing training modules and media materials, and supervising, monitoring and evaluating their implementation. I learnt a lot from all my colleagues in these agencies and departments and I thank them all. I would also like to thank innumerable other professional colleagues and friends and countless field staff, volunteers and activists whose natural communication skills and insights into human behaviour have taught me a lot.

I would like to thank Anil Ahuja for his help with the diagrams and artworks.

Last of all, I would like to thank my friend Harsh Sethi who encouraged me to write this book. Thanks are also due to Tejeshwar Singh and his colleagues at Sage Publications who agreed to publish the book.

Avik Ghosh

Introduction

The idea of this book occurred to me in the course of my professional association with large-scale state-led and/or donor-assisted development programmes. Over the years, I found that there was a paucity of documented experiences regarding the applications of communication technology in development. The thorough documentation of the Satellite Instructional Television Experiment (SITE) in 1975–76 remains a rare exception. The absence of any institutional memory in the relevant departments resulted in recurrence of faulty planning and unrealistic expectations. Development communication continued to be regarded as production of media materials and publicity in the media without much thought given to research and planning, identifying audience segments, developing messages and monitoring their implementation or assessing their effectiveness.

Having had the opportunity of working in several programmes in the social sector, I am able to recount and document the process of evolving a communication strategy and its implementation. The choice of appropriate technology, the issue of access to media, training of field personnel and evaluation of the communication materials are critical in a development programme. A descriptive account of past experience would give the reader an understanding of the practical issues that arise in the planning and implementation of communication programmes to bring about behaviour change.

The first part (Section I) of the book gives a historical background to the evolving nature of the application of communication technology in development. From the early thrust of expanding mass media reach and coverage to disseminate information and messages put out by the government, communication programmes have grown to focus on modifying behaviour. A combination of social and political consensus building through advocacy and mobilisation of the community, seeking their participation in a development programme, is used to address the difficulties that individuals and families encounter in adopting new behaviour. The development agenda-setting process has also evolved through representation of different interest groups concerned with human rights and conservation of the environment. Case studies of innovative applications of communication technology, both in India and abroad, are briefly described to allow the reader get a holistic perspective of the way the role of communication in human development has evolved over the decades.

The main content (Section II) of the book is divided into three parts—literacy, population and rural development. The focus is on recent experiences in these social sector programmes. Led by the government, major initiatives have been taken to apply communication technology in the design and implementation of the programmes. I use the term 'communication technology' in a broad sense to mean a whole system that includes setting objectives based on research, programme design and planning, application of hardware, preparation of materials, and professional management that is accountable. The case studies included in the book are based on my close association with these programmes. I do not claim that they are necessarily complete or comprehensive. Equally, there have been other impressive experiences in the application of communication technology in the social sector that have not been included because of my lack of sufficient acquaintance with these programmes.

A key feature of development programmes has been the recent effort to decentralise planning and implementation, with local communities being mobilised to take greater control and ownership of the programmes. The passage of the 73rd and 74th Amendments to the Constitution of India creating a third tier of governance institutions has resulted in the election of people's representatives to the different Panchayati Raj Institutions (PRIs) at the village, block and district levels. Apart from that, community-based and non-governmental organisations have been encouraged under these development programmes to inform and educate people, particularly the more socially and economically disadvantaged groups, regarding the many development schemes and programmes and enable them to benefit from them.

The adult literacy effort through the National Literacy Mission (NLM) in the early 1990s has been one of the more successful social mobilisation efforts with a massive volunteer participation, especially in the early years (1990–93). The evolution of the Total Literacy Campaign (TLC), with extensive participation by the local community at all levels, was based on the success of the Ernakulam district model. The multi-pronged communication strategy using mass media and *kalajathas* (local and traditional theatre performances) supported with interactive group meetings in villages created a people's movement for basic education in several districts where the TLC was successful. Interestingly, NLM also used radio and television for direct literacy instruction and these experiences are also documented in the book.

India's population programme had been focused on family planning to limit births and fertility control through a female sterilisation programme. The target-driven programme had come in for a lot of criticism from women's groups and the unsatisfactory and uncaring service delivery had been indicted by the National Family Health Survey (1992) as well. In keeping with the Programme of Action adopted at the International Conference on Population and Development (ICPD) in Cairo in 1994, India launched a new Reproductive and Child Health (RCH) Programme. This meant a sea-change in the programme design wherein informed choice, availability of reproductive health care services, gender sensitivity and client satisfaction were some of the key elements. Community Needs Assessment (CNA) and demand generation through creation of awareness regarding antenatal care (ANC) and institutional delivery were other thrust areas. This required a new training programme for the entire health service delivery personnel and a new orientation that inculcated a more responsive attitude. Developing interpersonal communication (IPC) skills was an important component of this training. Communication had to play an important role in bringing about behaviour change, both among service delivery staff and the client community, and the process of strategy development, training in communication skills, engaging private sector professionals and using research-based information in planning communication programmes were some of the new approaches tried under the RCH programme. The case study in the book documents much of this effort in some detail.

While the experience in adult literacy and population takes a national perspective, case studies from the states are highlighted in the area of rural development. One of the most significant experiences in recent years has been the large-scale training of elected members of the PRI to understand their new responsibilities and the functioning of PRIs. The Government of Karnataka's experience in using a satellite-based training system has been very interesting and successful. This technology is being extended to other states as well and the case study documented here is valuable in this context.

In rural sanitation, the innovative social marketing approach being successfully implemented in West Bengal shows that collaboration between PRIs and NGOs is possible. Effective implementation requires sensible planning, capacity building, management support and monitoring.

Poverty alleviation programmes usually have a capacity-building and empowerment component, in addition to funds disbursement to the identified beneficiaries to augment their income earning capacity. Madhya Pradesh, under the Indira Gandhi Garibi Hatao Yojana launched in 2001, gave a lot of emphasis to the communication component of the programme to ensure that empowerment and capacity building actually took place and dependence on state subsidy reduced. The case study documents the early experience in planning the communication component of this programme where the pressure of rapid implementation of the programme and disbursement of funds tended to neglect the issues and concerns of the people. Consequently, there was a danger of the programme not meeting its 'empowerment' objective, especially for vulnerable groups like women who are not drawn into the planning and implementation process.

Jharkhand presents a different picture of planning and implementation of a Joint Forest Management (JFM) programme for arresting the loss of forest cover and meeting the needs of the tribal communities living in these areas. There is a natural conflict of interest as the tribal communities find themselves economically impoverished, marginalised and alienated from the mainstream development process. Communication plays a critical role in establishing a dialogue between the tribal communities who feel that their traditional rights of access to the forests are being denied and the Jharkhand Forest and Environment Department (JFED) who have the responsibility for conserving the forests.

In the concluding chapter, I have attempted to flag a few issues and concerns that are critical in today's context of India's development.

Drawing on the experiences described in the book, I argue that the importance of communication in facilitating social change is not understood or appreciated sufficiently by the government. Application of communication technology in social and human development programmes is a complex task and requires a professional approach. Many of the present problems stem from deep-seated social norms and economic and social discrimination resulting in attitudes that are unresponsive to positive change. Service providers in government programmes often assert a misplaced authority, particularly with regard to disclosing information or being sensitive towards vulnerable groups. In a democratic country people have the right to information and their contribution is necessary for a realistic setting of the agenda for India's future development. This is the communication challenge and task before us.

Section I
HISTORICAL PERSPECTIVE

Communication and its Role in Development

Soon after the end of World War II the importance of mass communication media like newspapers, radio, cinema and television as accelerators in spreading ideas about progress and development began to be recognised. The successful dissemination of war propaganda through newspapers, film and radio showed that these mass media could mobilise people and opinions very quickly through their ability to spread the same messages to a very wide and large audience almost instantly.

The centrality of newspapers for the creation of a 'public sphere' where people recognise, share and discuss ideas and issues of common interest in a democracy had been acknowledged since the advent of the printing press, which was regarded as a harbinger of the modern, industrial age. Many scholars have argued that the availability of printed books created the individual's sense of identity and fostered the idea of early vernacular nationalism in Europe in the sixteenth century (McLuhan 1969). What print did so well, other (later) mass media like radio, film and television did equally well, if a little differently. The war broadcasts mesmerised the listeners with the feeling of intimate contact with leaders regarding patriotism and they were willing to contribute their bit for the victory of their country. Cinema with its allurement of visual reality captured vast audiences through a common simulated experience.

Based on the experiences of rapid economic development in Western countries, scholars recognised the role of mass media in the modernisation process—industrial development and urbanisation, expansion of markets, economic growth and socio-cultural well being. Many argued that the modern mass communication systems were agents of social change as well as indicators of economic and social progress. It was believed that the movement was always from an oral (or traditional) culture to a mass media-based modern society.

In the context of the economic development of developing countries, most scholars, predominantly Western, believed that neither modern science and technology nor modern institutions were sufficient to bring about change in a society whose people were believed to be basically traditional, uneducated and unscientific in their thinking and attitudes. The role of mass media systems was seen as critical in spreading the ideas of social mobility and economic progress. Wilbur Schramm noted:

> The task of mass media of information and the 'new media' of education is to speed and ease the long, slow social transformation required for economic development, and, in particular, to speed and smooth the task of mobilising human resources behind the national effort (Schramm 1964).

Daniel Lerner examined the correlation between the expansion of economic activity (being equated with development) and a set of modernising variables, chief among which were urbanisation, literacy, mass media use and democratic participation. A key finding was the strong correlation between the spread of literacy and mass media systems in an urban setting that led to the emergence of a nucleus of individuals who were highly motivated and adopted modernising influences within the Third World societies and spearheaded a climate of acceptance of change in these societies. Implicit in these formulations was the belief that the interaction of literacy and mass media was the means by which the vast majority of people in developing countries would 'eventually break free of their stupefying bonds of traditionalism, heralding, as it were, the *passing of traditional society*' (Lerner 1958).

THE DOMINANT PARADIGM

With hindsight, it appears that it was rather simplistically believed that the task of modernising the Third World was just a matter of determining appropriate technological innovations promising high pay-offs and arrange to have them diffused to targeted beneficiaries. The inadequate industrial infrastructure and 'backward' cultures of these nations were perceived to be the major obstacles to progress and modernisation, which were regarded to be the panacea for the poverty and miserable conditions prevalent in these countries. The long history of colonial exploitation and the oppressive social–structural conditions were not given enough attention in view of the optimism based on the positive experience in developed countries. Thus, Third World countries were considered to be in the same stage as the European nations before the onset of the industrial age. It was consequently believed that, by retracing the path of the European countries, the Third World could be speeded through the different stages of development at a rate even faster than the advanced countries themselves experienced. It is this orientation that came to be known as the *dominant paradigm*.[1]

The trajectory of progress in the dominant paradigm envisaged a transition from traditionalism to modernisation in a unidirectional way. This mindset led to an approach to communication for development that was 'vertical in nature, authority-based, top-down, expert-driven, non-negotiable, well-intentioned, and exhorting people to adopt new attitudes and behaviours that held promise of a much better future'(Ascroft and Masilela 1989). For the newly independent countries emerging from long years of colonial rule, the extensive use of mass media was prescribed as a way to quickly reach the messages of modernity to vast numbers of people who were still living in rural isolation. Development involved not simply the transfer of technology, but also the transmission of ideas, knowledge and skills to make possible the successful adoption of innovations. Wilbur Schramm, in his book *Mass Media and National Development,* remarked that 'knowledge is better than ignorance; health is better than disease; to eat is better than to be hungry; a comfortable standard of living is better than poverty; and to participate actively in one's nation is better than to be isolated from it' (Schramm 1964). This information transmission model (see Box 1.1) became the credo of development aid and financial support for expansion of the broadcasting systems in developing countries.

> **Box 1.1**
> **The Information Model[2]**
>
> The information transmission model assumed that the communication process involved the unidirectional flow of information from a centralised source (sender) using an appropriate mass media (channel) to communicate the relevant information (message) to the widely dispersed audience (receiver) to achieve the desired objective (effect). Communication was conceptualised as a simple, mechanistic, process of message transmission. In the post-World War II period, the SMCR model became popular with communication for development professionals:
>
> Source (S) > Message (M) > Channel (C) > Receiver (R)
>
> Adoption was defined by Everett Rogers as the process through which individuals arrived at the decision to adopt or reject the innovation from the time he first became aware of it. The five stages were: awareness, interest, evaluation, trial and adoption. Rogers later described them as knowledge, persuasion, decision, implementation and confirmation.[3]
>
> While some researchers believed that communication acted as the harbinger of new ideas from outside, others felt that communication (as transmission of new ideas and information) helped to smooth the transition from a traditional to a modern community. If only attempts were made to open up a traditional society to modernising influences through communication, the new information available to the people at the top and its eventual and autonomous 'trickle down' to others in the lower reaches of the hierarchy would increase 'empathy', open up new opportunities, and lead to a general breakdown of the traditional society.
>
> The strength of the mass media lay in their one-way, top-down and simultaneous wide dissemination. And, since the elites in every nation were required to modernise others in the population, the control of the prestigious mass media by them served their economic and political interests. They were considered as *magic multipliers* of development benefits in Third World nations. Administrators, researchers and fieldworkers sincerely believed in the great power of the mass media as harbingers of modernising influences. A spirit of optimism characterised the early period of development history of the newly independent nations in the 1950s and 1960s.[4]

THE INDIAN CONTEXT OF COMMUNICATION FOR DEVELOPMENT

It is interesting to note that even before Independence there had been a section of the colonial administration that saw the potential of radio broadcasting as a means of reaching the large mass of the rural population with information regarding agriculture and other relevant revenue and administrative issues.[5] While this approach did not find favour with the government at that time, after the Government of India Act of 1935 the debate centred round the issue of provincial autonomy and whether radio broadcasting should be a central responsibility or handled by the provincial governments. The Indian nationalists were suspicious of the government's initiative and regarded broadcasting as a tool of British propaganda; the Government of India, on the other hand, was equally reluctant to hand over the control of broadcasting to provincial governments in order to contain the 'fissiparous tendencies' in the different regions. That, and the later but more urgent compulsion to counter the war propaganda from Japan and Germany, sealed the decision of broadcasting being under the strict control of the Government of India. However, until Independence, broadcasting was limited to mainly news and current affairs with some modest but significant amount of music and other cultural programmes. It was still a new medium and its uses were yet to be explored. Despite the enthusiasm of the broadcasters, the number of radio receiver sets in use in the whole country before 1947 was only 85,000. While this figure rose substantially after Independence, it was still relatively low because of the high cost of radio receivers and their dependence on electricity or heavy consumption of batteries. It was only after the transistor revolution in the early 1960s that radio became a mass medium and spread to rural areas.

With Independence, the Indian leadership felt the need to unite the country and create a national identity. This was especially urgent in the aftermath of Partition, Gandhiji's assassination, influx of refugees and other divisive tendencies. Radio was an important instrument that the Central government could control and use to give news and information to the people and elicit their support for its nation-building effort. The number of radio stations rapidly expanded from 5 to 25 between 1947 and 1951.[6]

India was launching itself as a modern secular and democratic nation and the five-year plan was a major instrument for achieving rapid economic development through public sector investments in industrialisation. This was in consonance with the international thinking on modernisation and represented the dominant paradigm. While Gandhi held that needs and wants must be limited and favoured decentralisation and strengthening the autonomy and self-sufficiency of the village economy, there was no serious consideration of this approach at that time. As Sukhamoy Chakravarty has observed, 'While the Gandhian approach has received a certain measure of support in the recent writings of ecologists and ecologically-minded economists, in the early fifties such positions appeared to lack any substantive theoretical foundation' (Chakravarty 1987). The debate at that time was between the relative merits of free market capitalism and Soviet-type socialism and the possibilities of the model of a 'mixed economy' as adopted by India, but all placed their faith in increased commodity production as a means to attain a better standard of living rather than delimiting wants and needs as advocated by Gandhi.

It was believed, in consonance with the international thinking on the role of mass media in national development, that reaching messages (new ideas and information) to the people was sufficient to bring about adoption of new behaviour and practices that were necessary in the modernisation process. This was based on the importance given to mass media in national development, with large sections of the population needing information about government programmes, economic opportunities and developing a sense of national identity.

In the First Five-Year Plan document in 1951, the government defined the role of communication in development thus:

> A widespread understanding of the Plan is an essential stage in its fulfilment.... An understanding of the priorities of the Plan will enable each person to relate his or her role to the larger purposes of the nation as a whole.... All available methods of communication have to be developed and the people approached through the written and spoken word no less than through radio, film, song and drama (Government of India 1953).

At the time of Independence, India had only six radio stations. In order to foster the sense of national identity, yet respecting the country's regional, cultural and linguistic diversity, radio stations with studios and lower power transmitters were set up in each state to begin with, and then expanded to cover each distinct linguistic region in the country. The five-year plans gave substantial emphasis to the radio network since it was the Central government's main vehicle of communication for reaching information to the people. News was important and during and immediately after national calamities like floods and cyclones, radio was often the only link between the government and the people.

RURAL BROADCASTING—RADIO RURAL FORUMS

Rural broadcasting was a significant part of the expansion plan of All India Radio (AIR) and community listening sets were installed in villages on a cost-sharing basis by the Central and state governments. The system was cumbersome as the radio sets had to be maintained by the state governments with batteries requiring frequent replacement in the absence of electricity.

An experimental project establishing Radio Rural Forums was conducted between February and April 1956 under a collaborative project with UNESCO. The idea was to establish a two-way communication link between village audiences and the programme organisers in the radio station. Each programme covered a subject of interest to rural audiences for 20 minutes and the remaining 10 minutes were left to answer listeners' questions. In all, 20 programmes were broadcast during the two-month period. A radio rural forum was set up in 150 villages around Poona (now Pune) with 20 persons in each village selected as members of the forum. Each forum had a chairman and secretary to conduct its meetings. Forum members listened to the specific rural programmes that were broadcast and later discussed the topics amongst themselves. If there were doubts or queries, these were jotted down by the secretary and mailed to the radio station. In the following programme, the programme organiser in consultation with specialists addressed these questions.

The experimental parameters of the project—information gain, audience interest and participation in the programme, role of the forum as a new, dynamic village institution—were evaluated at the end of the project and the findings were very encouraging.[7] Not only was there substantial increase in knowledge among the forum members as compared to the others, the forums also became active decision-making bodies capable of speeding up common pursuits more successfully than the elected panchayats. In other words, radio rural forums demonstrated the potential of creating a dynamic spearhead in villages to encourage adoption of new attitudes and practices.

Based on the success of the Radio Rural Forum experiment in Poona, the coverage was extended to all radio stations broadcasting rural programmes. Community listening sets were distributed and radio rural forums began to be set up in villages surrounding the radio stations. This was a hasty step with insufficient planning. By 1964 it was claimed that over 7,500 radio rural forums were in place averaging 250 forums for each of the 30 radio stations then engaged in rural broadcasting. The maintenance of radio sets and irregular supply of batteries posed a continuing problem for regular listening, the initial enthusiasm of the Poona group did not get extended to other stations and the coordination with state government departments of agriculture and rural development faltered. By the mid-1960s rural radio forums had become a failure. Rural broadcasting continued in one form or the other and listeners' questions were answered in these programmes but it did not have the same spirit and dynamism of the initial experiment. The next major initiative in using radio for rural development had to wait till the transistor revolution had taken root and the Green Revolution—introduction of high-yielding varieties of seeds, intensive use of chemical fertilisers and pesticides in agriculture—required a qualitative change in Indian agricultural practices. This needed a major communication intervention through agricultural extension agents and use of radio and film publicity.

FOSTERING A NATIONAL IDENTITY

The first decade of the Indian republic was a time of optimism and hope. It was a time for throwing away the baggage of colonial rule and marching ahead with the agenda of building a secular, democratic and socialist republic firmly on the path of modernisation. The various provinces had to be restructured on a linguistic principle and this task was executed mainly during the period between the late 1950s and mid-1960s. It created political units that became coterminous with large linguistic identities.

Identification of the boundaries of a state on the basis of one language which was culturally predominant and also numerically preponderant in that region or province, and the recognition of that language as the official language of the state created a strong and stable cultural base for the political-linguistic identities in the country. In the process, however, the other smaller languages and cultures in the regions became marginalised. This problem was especially acute in the large Hindi-speaking states where a variety of languages like Maithili, Bhojpuri, Bundeli, Marwari and others were submerged in the mainstream 'official' Hindi of administration and formal education. The 'non-official' languages in the linguistic states do indeed survive today, but more as 'spoken languages' or 'dialects' or media of expression of 'folk cultures'. Their role in formal education and in the administration of these states has been almost erased.

Unlike the Hindi-speaking states where the earlier languages (variants of Hindi) of the area have lost their importance, other regional languages have become the vehicle of mass literacy, formal education and generally of public communication in the state. As a result, the regional languages themselves have undergone significant transformation. The socio-cultural groups, which earlier had little or no access to formal education, began to acquire significant levels of education through the medium of an officially recognised regional language.[8]

While language was the criterion for the reorganisation of the states, the federal structure of India needed careful nurturing, what with problems of poverty, low literacy and low economic productivity looming large in most parts of India. In addition there was the 'wound' of Partition and the influx of refugees. Daunting problems, no doubt, for a nation newly independent. However, there was buoyancy in the faith that these problems would be overcome under the charismatic leadership of Nehru and the Congress party.

The rude shock of the Indo-China war of 1962 and the humiliating defeat suffered by India left an indelible impression on the national consciousness. The failure of the development effort to arrest poverty and improve literacy, lack of communication infrastructure, particularly in the border areas, the burgeoning population and continued dependence on food imports hastened the need for expanding the government's communication infrastructure. The Directorate of Field Publicity was set up under the Ministry of Information and Broadcasting as regional information units with mobile cinema vans that showed films on government programmes and furthered a sense of national unity and belonging.

Information films got a fillip at this time with Films Division documentaries and weekly newsreels being compulsorily screened in all cinema halls at every show. By the mid-1960s films on India's defence preparedness and the role of ordinary citizens in times of war or conflict with neighbours were being given wide publicity. Radio transmitters were also expanded to cover border areas more effectively.

Agriculture and family planning were two other areas that became important in that period for information dissemination through the mass media. The alarming growth of population alerted the international community to recommend urgent measures for population control as the investments in economic development appeared to be negated by population growth. Poverty and deprivation were increasing and the expansion of education and health facilities could not keep pace with the growth in population.

USE OF FILM AND RADIO IN THE GREEN REVOLUTION

The problem of food shortage coupled with low agricultural productivity also required urgent attention and necessitated the adoption of new agricultural practices dependent on high-yielding varieties of seeds and intensive use of chemical fertilisers and pesticides. The Green Revolution, as it came to be called later, required the promotion of capital-intensive and increasingly mechanised farming practices. Indian farmers had to be informed and educated in these new agricultural practices. An extensive information dissemination campaign was launched with several agricultural universities spearheading the campaign through their extension services and establishment of Krishi Vigyan Kendras. Films and audiovisual materials were distributed widely, with regular radio broadcasts being an essential feature of this campaign. Consistent with the research finding that innovation and change required mass media and literacy to act in concert, a Farmers' Functional Literacy Programme was started in the mid-1960s that would give agricultural information through films and radio programmes combined with face-to-face interaction with extension agents and literacy workers. The emphasis was on functional literacy since familiarity with the new seeds, fertilisers and cultivation procedures was thought to be more necessary than mere cognition of the alphabet. Unfortunately, the adult literacy component was not given due attention and the programme was restricted to publicising the new agricultural practices through the government-owned media of radio and film.

Subsequent research showed that the main beneficiaries of the Green Revolution were the economically better-off and educated farmers who took advantage of the various incentives offered by the government. Also, the main impact of the Green Revolution was limited to areas that had assured rainfall and irrigation facilities. Where land reforms had not taken place and social discrimination and poverty were acute, the Green Revolution had little impact. The hope that the benefits derived by the 'early adopters' would percolate to others did not bear fruit in the Indian context. However, the Green Revolution initiative did see the expansion of agricultural extension services relying substantially on mass media (film and radio) in order to reach the message of new agricultural practices to rural farmers. Indeed, the significant use of these media has been one of the key contributory factors of the Green Revolution, according to some observers.

FAMILY PLANNING COMMUNICATION

The Family Planning Communication Programme got a big push with the creation of the Mass Education and Media (MEM) Division within the newly established Department of Family Planning in 1966. The adoption of the 'extension education' approach, with fieldworkers reaching out to people to motivate them to adopt family planning practices, meant the creation of a network of field personnel at the national, state, district and block levels, following the pattern of the agricultural extension services already established by then. Radio and film also began to be used in a limited way. Films, owing to their powerful impact as an audiovisual medium, were also seen as a major vehicle of communication and the district units of the MEM Division were equipped with audiovisual vans for exhibiting motivational films. It was in this period that a strategy for communication and a pinpointed, clear and specific message to the family were articulated for the first time. The *Red Triangle* symbol for family planning was developed and slogans propagating 'two or three children—enough' and the small family norm began being used as a mass campaign.[9]

This approach did not take into consideration the socio-economic conditions of specific audiences and the producers based in big cities had hardly any experience of rural reality. Centralised production of films, which were then dubbed in different languages, had its own drawbacks as audiences found it difficult to identify with the images and cultural context. Very often, the messages promoted through the films had no connection to the services available on the ground. The rational argument in favour of small families did not properly address the situation of rural families where infant mortality was high and additional hands for family labour meant additional income. The family planning extension staff in primary health centres had limited facilities and little control over the exhibition of films or training and preparation to use them in their work.

This lack of coordination between a centralised system of preparation and distribution of audiovisual media materials and the hard reality of the audiences in villages and the dissonance between the projected promise of services and what was available locally created a credibility gap regarding the mass media projections of the government (see Box 1.2).

SATELLITE INSTRUCTIONAL TELEVISION EXPERIMENT (SITE)

In this era of gloom there remained one shining example of a fairly successful and well-planned and executed experimental communication project to foster economic and social development. The

> **Box 1.2**
> **The Emergency—Loss of Credibility**
>
> Unfortunately, the Emergency period (1975–76) with its mass sterilisation camps (with some element of coercion) and the imposition of press censorship that suppressed negative feedback, created a horror of family planning in the public mind. While targets may have been achieved in that period, the use of force, employed by many countries to bring down population growth rate, did not work in India and planning one's family remained the voluntary choice of married couples. The coercive family planning measures imposed among the poor in parts of North India through mass male sterilisation camps created a wave of antipathy towards the government in power. This, along with press censorship and careful scrutiny of radio and television broadcasts that were used only for government propaganda, eroded the credibility of the government-controlled media (radio, films and television) completely. When the next general elections came around in 1977 the Congress party was defeated and the first non-Congress government was formed in New Delhi. Misuse of a centralised mass media system proved to be as much a cause of the government's undoing as press censorship and the arrest of opposition leaders. Democracy had taken root and the people voted the Congress out.

Satellite Instructional Television Experiment (SITE) was a landmark global experience in the use of satellite broadcasting to remote rural areas on issues of agriculture and other development and welfare concerns including health and family welfare.[10] Vikram Sarabhai, the pioneer scientist and champion of SITE, defined the objectives of the experiment in the following words:

> If India wants to reduce the overwhelming attraction to cities, enrich cultural life, integrate the country by exposing one part to the cultures of the other parts, involve people in the programme of rural, economic and social development, then the best thing is to have TV via satellite (cited in Chatterji 1987).

SITE amply demonstrated that sophisticated broadcasting systems could be used in the context of rural communication quite effectively and that it was quite possible to organise and manage the maintenance of Direct Reception Sets (DRS) for TV viewing in rural areas. It also showed that programmes based on familiarity with the socio–economic and cultural background of audiences and representing their felt needs had greater likelihood of evoking involvement and interest. If the telecast was supported by coordinated follow-up by the field staff, the development objectives of the telecast were more likely to be met. In other words, television, or for that matter any mass medium, did not work by itself. Support from extension workers through Interpersonal Communication (IPC) was necessary. However, the results of SITE also showed that the mass media were not a 'magic wand'. Poverty and illiteracy were basic problems and benefits of new information and technical inputs or services could be availed only by those who had crossed that threshold.

NOTES

1. For a clear enunciation of the 'dominant paradigm' regarding the role of communication and its relationship with the modernisation process initiated in developing countries, see Melkote (1991).
2. Much has been written on this model that for decades has been regarded as a standard prescription for development communication. This uni-directional flow of communication was conceptualised in the early 1950s based on the Lasswell formulation: 'Who says what in which channel to whom with what effect' (see Lasswell 1948). Srinivas Melkote gives a fairly comprehensive review of the history of the development of this model in his book cited earlier.
3. See Rogers (1962). This has continued to influence theory and practice of development and communication practitioners regarding the steps by which new ideas get adopted. Rogers himself has refined his ideas over time based on further research and experience of developing countries and non-Western cultures.
4. The research of Lakshmana Rao in India seemed to support the thesis that exposure to new ideas through mass media had a definite influence on adopting new technology and modern practices (see Rao).
5. For a review of the early history of Indian broadcasting (and much else) the book by David Page and William Crawley is excellent (see Page and Crawley 2001).

6. The figures on the number of radio sets and other factual details are taken from the fairly comprehensive history of broadcasting in India by P.C. Chatterji (1987).
7. The findings of the study are described in Chatterji's book (1987). The evaluation study was conducted by a group of social scientists under the aegis of UNESCO.
8. I have also written about the interplay of language, education and media elsewhere (see Daswani 2001).
9. For a detailed review of the family planning communication programme in India, particularly the early history, see Kakar (1987).
10. Among many other writings on the subject, a summary description and evaluation of the impact of SITE may be found in the technical report by Binod C. Agrawal (1978).

Search for an Alternative Development Paradigm

By the mid-1960s, the assumptions of the dominant paradigm based on the experience of the developed countries of Western Europe and North America began to be questioned as being valid for the path to be followed by the newly independent countries of Asia, Africa and Latin America. In fact, the underdevelopment of Third World nations was regarded by many scholars as a direct result of colonial exploitation that led to the rapid economic development of European and North American countries. While these countries prospered on the easy availability of cheap raw materials for their growing industries and an expanding market for their machine-made products, the colonies stagnated owing to the decline of their traditional crafts and household industry and their inability to compete with imported machine-made goods. Underdevelopment in the poor countries, it was argued, was and still is generated by the very same historical process that generated wealth for the developed countries: the development of capitalism itself (see Frank 1979).

The emphasis on the emergence of the 'modern' person or individual as being the instigator of change was again based on an inadequate understanding/appreciation of the social structures in developing countries, where submitting to the collective will and dominance of group identity played a more important part. Breaking these ties required much more patient effort than mere literacy and the induction of new technology and innovative practices. The social structures operated at several levels and influenced economic and political decisions at local, regional, national and international levels. In the Third World countries, the ruling elite used their relative wealth and power to their advantage and limited the percolation of benefits of economic development to the large mass of people in these countries.

Charged with the ideology of 'modernisation', highly motivated individuals effectively sustained the existing unequal structural relationships in developing societies. Without distributive justice being made an objective of development, the economic endeavour only perpetuated the skewed distribution of income, wealth and power. Rise in Gross National Product(GNP) or per capita income as a measure of 'development' did not necessarily improve the standard of living of all people. Very often, the gap between the rich and poor in these countries grew much worse.

Disillusionment with the dominant paradigm's failure to deliver the 'promise of plenty' set in by the early 1960s as several developing countries faced various economic and political crises—growing poverty and unemployment, spread of disease, ill health and high infant mortality, insufficient expansion of basic education and skills development—apart from the continuing occurrence of natural calamities like floods and droughts. The governments in the newly-independent developing countries, with their agenda of modernisation, failed to deliver on their promises and the successful model based on the Western experience largely benefited the ruling elite and educated middle class in these countries.

In short, the dominant paradigm was found wanting in several respects. Defining development only in terms of quantifiable measures like GNP and per capita income was inadequate in view of the gross inequality in the distribution of income and wealth in these countries. Pursuing capital-intensive strategies to create an industrial base thereby neglecting labour-intensive strategies (in labour surplus conditions) was also criticised by scholars. The presumption that a country could focus only on its internal constraints and ignore the influence of larger forces outside its control was fallacious. Finally, the top-down approach to planning and development, with well-meaning paternalistic bureaucracies taking decisions on people's behalf, failed to reflect the new concern for self-reliance and need for self-expression through popular democratic participation in development activities.

As a consequence of the propagation of new values and attitudes through the mass media and other channels, and the increasing influx of the market economy, the traditional support structures available to the poor in villages through the joint family, caste and kinship network or the religious institutions started to break down. There were also adequate alternative social welfare systems provided by the State.

Questions began to be asked: What is development and for whom? Who determines the priorities and who reaps the benefits? The ethical dimension of the development process came into focus and scholars felt that development should have a more 'humane' approach that reduced human suffering rathar than increased it. Thus development was redefined with this basic needs approach that highlighted the elimination of some of the worst aspects of absolute poverty. New indicators of development included provision of adequate food and clean drinking water, decent shelter, basic education, security of livelihood, adequate transport, and opportunity for people's participation in decision-making, with emphasis on upholding a person's dignity and self-respect.

In the dominant paradigm mass media had an important role in reaching information from the centre (national capital) to the periphery (rural communities) and persuading people to adopt new behaviours and practices. It was presumed under the Source–Message–Channel–Receiver (SMCR) model that the receiver would aspire for 'the good things of life' as projected in the media and therefore work harder to increase income and improve life circumstances. It ignored the structural constraints prevalent in developing countries and did not regard them as a barrier.

In the top-down communication model implicit in SMCR, the receiver is a passive recipient and the control over what messages are to be communicated vests with the centralised information and broadcasting organisations directly under the government. In many countries, including India, these structures were inherited from colonial times. There was also a presumption about literacy and access to media channels, which were belied by the reality on the ground. Those living in rural areas, without electricity and with poor connectivity by road and rail, had limited or no opportunity to visit the cinema, watch television or listen to the radio. Till the advent of transistor radio sets, the dispersion of radio listening was very limited and even afterwards, it was restricted to those with the purchasing capacity and resources to replenish batteries regularly.

It is in this context that the traditional theatre and other folk media forms came to be seen as possible vehicles for communicating messages to the rural people. The rich cultural tradition in different parts of India that was rooted in the language and culture of the local people appeared to be convenient and appropriate. The Government of India established the Song and Drama Division and many state governments followed with their own folk entertainment cultural troupes. Many regional centres and several troupes and folk performers of a variety of traditional media forms were registered with the Song and Drama Division and gave performances on development themes before rural audiences.

UNESCO promoted the use of traditional media actively as a means to preserve the cultural heritage and livelihood of a large number of traditional performing families.[1] They were being edged out by the popularity of the Hindi cinema and film music. The other reason for promoting the use of traditional media was the familiarity of these artists with the local milieu and their ability to weave in the new messages of development into their performance. Also, these performers were itinerant troupes that moved from village to village and performed at traditional fairs and festivals. Because of their familiarity with the audience, a greater credibility and acceptability of the new messages was presumed. However, experience showed that the performers did not always understand or accept the development messages that they were promoting for the government. Since these media were used for narrating locally well-known popular myths and legends, integrating new messages was neither easy nor acceptable to the audiences. Some traditional forms allowed easier interpolation of new messages during the performance. Where the script and messages were tailored to suit the local situation, they worked better. Within the framework of a centralised organisational structure, the messages were generated from a distance and the local performers had little or no flexibility or capacity to modify them.

In the search for an alternative paradigm of development, communication also had to be viewed differently. The dominant paradigm projected its messages through the mass media in a transmission/persuasion mode and was consequently alien to the people culturally. Also, mass media materials were produced by urban and educated producers and inherently had a pro-literacy bias in terms of using technical language and symbols unfamiliar to the rural people. The emphasis was on reaching the message and communication was therefore dominated by technology and expansion of hardware availability—more radio and TV stations and more powerful transmitters. Producers were influenced by the standards of technical quality set by broadcast engineers and programmes tended to be urban-centred and studio-based with little connection to the lives of rural audiences.

The use of mass media was questioned and traditional media gained acceptance as a legitimate vehicle of rural communication complementing the mass media. The other search was for an alternative to the SMCR model of transmission/persuasion that would engage in a dialogue with the audience. In other words, the *process* of communication had to be given adequate emphasis. Therefore, face-to-face communication and small group communication media gained ascendancy during this period.

Several international agencies like the Food and Agriculture Organisation (FAO), United Nations Children's Fund (UNICEF), World Health Organisation (WHO) and others that were active in the field of human welfare and development focused attention on the need to build the capacity of frontline workers—the agricultural extension worker, the village health guide, the auxiliary nurse/midwife—in interpersonal communication skills. Visual aids like flannel graphs, flipcharts, posters, filmstrips and audiocassettes were some of the support materials provided to these workers to organise small group meetings in villages. The problem of distributing these centrally designed materials and then producing them in local languages was a problem in itself. Also, this measure was only a substitute for the mass media as there was little effort to engage the audience in thinking or articulating their needs and aspirations or finding their own solutions.

It is in this situation that the role of voluntary agencies gained importance in India. In the Indian tradition, charitable and voluntary community work has always been accepted and respected. During the freedom movement, Mahatma Gandhi inspired a range of selfless and dedicated workers who chose the path of working for the development of rural people. By the turn of the 1970s, as a consequence of growing educated unemployment and student unrest, political upheaval, increasing rural poverty and the Naxalbari uprising, widespread dissatisfaction among many young educated people was evident. Perhaps this was also the time, after the liberation of Bangladesh, that substantial funds from many private international charities became available to fund voluntary agencies in India for relief, rehabilitation and reconstruction work.

Beginning with mitigating the suffering wrought by hunger and natural calamities, these agencies quickly extended their work to include informal education and creating awareness among the rural poor about the government's development schemes and their rights and entitlements. The crucial factor in favour of the 'development workers' of the voluntary agency was their willingness to listen to the people and work with them. They worked in the fields of adult education, community health services in under-served areas, developing design and marketing links for local skills in traditional crafts and other occupations, and as facilitators for skills development and local resource management.

This approach of voluntary agencies echoed the disenchantment with big industry and mass production in the face of wanton destruction of the environment and a move towards intermediate (or appropriate) technology that had Schumacher's 'Small is Beautiful'[2] as its credo. It demonstrated that a more 'humane' approach to development with people's participation *was* possible, albeit on a small or even tiny scale of operation. The key was the involvement of the people in the development process, something that was missing in the dominant paradigm. Reaching across to people and engaging with them, these agencies demonstrated the richness of local experience and creativity in finding solutions to local problems without large investments or government intervention. In communication terms, it meant a 'dialogical approach', which was different from the transmission/persuasion approach of the dominant paradigm.

Working among poor peasants in Latin America, Paulo Freire (1971) had started a new way of approaching adult literacy programmes. He argued that literacy was not merely learning 'to read the word but to understand the world'. Traditionally, in the colonial context, it was presumed that illiterate peasants were ignorant and education meant depositing information and knowledge into an empty mind. Freire called this the banking concept of education. He argued that this led to a 'domestication' of the poor with an uncritical acceptance of authority. If change had to be brought about in the lives of the poor, understanding the causes of their suffering through critical reflection was necessary. According to Freire, for the poor, literacy had to be an engagement in a process of becoming aware of the causes of their misery. It was a powerful tool for them to obtain their rights and entitlements and thereby bringing about a social transformation. This became a refreshingly new approach to communication as well, as it laid emphasis on the sense of respect and equality between the peasant and the adult educator, or social animator as he called them.

In this new approach to development that sought a more direct participation of people in the development process, there was a greater emphasis on dialogue and the symbolic representation of the communication process looked more like this:[3]

I (interlocutor) ◄──────► Media ◄──────► I (interlocutor)

The interlocutors on both sides use media—interpersonal, folk, small-group as well as mass media as an active engagement with one another with the objective of sharing experience, knowledge and skills in order to evolve to a higher plane of awareness through this process. Mutual respect and trust is established through this process and a commitment made to develop themselves together. Freire used the term 'conscientisation' (in contrast to 'domestication') to express the feeling of awakening and breaking free from the 'culture of silence' and passive acceptance in which the large majority of the rural poor were immersed. The social animators in the Freirian scheme of things were not literacy instructors but activists who participated in the churning of the minds of the poor and illiterate learners as they learnt the alphabet (using a different andragogic method) and absorbed the power of the written word. It was an entry into the world of the powerful and literacy became a tool for them to claim their legitimate share in the nation's development. This foundational principle of a literacy programme placed emphasis on the empowering potential of a communication process. The engagement between the interlocutors was on the basis of equality and not on a hierarchy of 'sender' and 'receiver' where the control lay with the sender of messages (in most cases, the government or those in authority).

It was believed that communication hastens social change. This remained a fundamental premise in the development process. In the earlier dominant paradigm, the assumption was that transmission of the right messages, as defined by the government and the ruling elite, would bring about social change through the adoption of new behaviours and practices by the people. As it did not yield the desired results, the entire paradigm of modernisation being imposed from above was questioned. The large-scale participation of people was seen to be necessary to define the very objectives of development and satisfying the basic needs of people. Therefore, communication's role changed to bringing about this change in the power relations in society so that genuine development of people could take place.

What were the guiding principles of any communication activity in this perspective? To bridge the gaps in knowledge between the government in urban centres and the people in rural areas, channels of communication were necessary to create a dialogue between the government and people and between different rural communities.[4]

There is no doubt that informing people about government policy and programmes was necessary and a forum for discussion with the people had to be created. Providing training and stimulus appropriate to local needs and conditions was also important. At the same time, it was equally important for the government to know what the development priorities at the local level were and what effect—beneficial or destructive—the development projects initiated by the government were having on the people in the rural areas. This would be regarded as a feedback mechanism or 'bottom-up' communication.

Finally, it was necessary to link the different rural communities together. This would be regarded as 'horizontal' communication. Through this process of connecting the relatively isolated rural communities a collective could be formed and concerted actions and alternative plans and viewpoints expressed and communicated or understood. It was through these efforts that networks and movements developed that then began to impact on the planning and development process as a whole.

It is against this perspective of an alternative approach to communication and development that we shall examine a few case studies and other experiences in the 1970s and early 1980s in India and elsewhere.

NOTES

1. An international workshop on the 'Role of Folk and Traditional Media in Development' was held in New Delhi in 1972–73 and the discussions centred on the importance of preservation of these folk forms as part of the cultural heritage as well as using them to meet the new communication challenges facing developing countries.
2. The critique of large industrial units and promotion of intermediate and/or appropriate technology was the alternative being successfully

demonstrated in several countries. Inspiration was provided by the writings of Schumacher and others (see Schumacher 1973).
3. The graphic representation of the interlocutor–media–interlocutor dialogical model was presented by Manuel Calvelo Rios of CESPAC, Peru, at a workshop in New Delhi in 1981 in the context of its work in Peru (discussed later).
4. A succinct presentation of this approach is given by Simon Bright (1981).

3
Alternative to Broadcasting: Some International Experiences

The radical alternative 'pedagogy of the oppressed' propounded by Paulo Freire to give voice to the oppressed poor and help them articulate and express themselves, influenced the search for alternative approaches to the use of audiovisual media as well. Freire's 'conscientisation' process involved looking at visual symbols/pictures of everyday life—peasants labouring in the field, the landowner's mansion and the peasant's shack/hovel—by which the poor were able to understand and revalue their life, learn new communication skills like literacy, and thereby control their environment better. The earliest example of a similar technique to give expression to the felt needs of the poor and present their articulations was first tried by a team from the National Film Board of Canada in the Fogo Island off the coast of Newfoundland.[1] The film crew recorded the views of the local community on the Canadian government's efforts to resettle them on the mainland. These views were then played back to different communities on the island and the films became a means of dialogue between the different communities and helped develop a consensus among them. The resulting film, which was made with the active participation of the islanders, was then sent to the Central government. This initiated a series of films that were screened back and forth and led to the decision of the government to allow the island community to develop the island under the control and direction of the local community.

The Canadian Film Board's *Challenge for Change* film series (that went beyond the Fogo Island experience) became a model for the later use of video as a participatory tool for development to present the viewpoint of the poor in a language and manner defined by them. Internationally, there were many experiences in using radio, audio, film and video in a similar way. Several international agencies supported the design and execution of such projects in developing countries. The Tanzania Year 16 (TY 16) project and the CESPAC (translated as Audiovisual Teaching Services Training Centre) project in Peru deserve special mention as examples of different approaches towards the use of media in rural development.

In the TY 16 project,[2] a three-member production team lived with villagers in a few Ujama villages in Tanzania for a period of six months, recording the process of economic and political reconstruction initiated by Tanzania's socialist government under the leadership of Julius Nyerere. They used portable video equipment to record public meetings and private discussions and played them back to the community. This process of reflecting back to the community their own views opened up discussion on issues that were not talked about otherwise. Specific criticisms of government programmes were shown to the authorities to encourage corrective action. The media became an objectifying 'mirror' of people's experience and enabled them to challenge the existing social, economic and political structures. Ordinarily, people would not have the courage to challenge the incompetence or corruption of government officials or point to the collusion between them and the powerful people in the village itself. The video recording enabled people to articulate their reality and critically perceive their condition when their own expressions were played back to them. This was possible because the new channel of communication was not controlled by the powerful in the village. Because the team members were outsiders, they were able to take the tapes and play them back to senior officials without interference from the local hierarchy. Since the TY 16 had the approval of the government, which was initiating a process of rapid social transformation, this experiment in using video as a social mirror enabled people to re-evaluate their own circumstances and initiate positive action. Their ideas and articulations were treated with respect by the authorities, thus bridging the gap between the people and the authorities. The production team acted as a 'facilitator' and worked with the people to present their world as the people themselves saw it. This was vastly different from the traditional broadcasting model where programmes were designed and prepared on the people and their problems but the depiction of their reality was as conceived by outsiders.

Video for Farmer's Training in Peru

Peru is a country divided into three distinct geographical and ecological areas consisting of the coastal region, the Andean mountain region with isolated villages and poor roads, and the tropical forest zone of the Amazon basin. The use of media thus had to take into account the vast regional variations in environment, language and culture. Peru inherited a broadcasting system with a network of radio and television stations which were mostly commercial, devoting much of their time to promote multinational brands and products. The contribution of the broadcasting system to national development was insignificant.

In 1968 the military revolutionary government of General Velasco transferred 30 per cent of land to peasant cooperatives. This was a major structural change effected towards improving the livelihoods of small peasants (*campesinos*). However, these 1,500 cooperatives along with the 4,000 native communities needed training in improving agricultural production. CENCIRA (translated as National Centre for Land Reform Education and Research) was established in 1970 to meet these training needs and the educational process was based upon a dialogue between the rural people and 'experts', mediated by the use of locally based video production units. CESPAC was established in 1974 and funded through FAO/UNDP and Dutch aid to provide support to CENCIRA's training activities.

CESPAC set up six regional production centres located in all the three main geographical areas of Peru in order to be responsive to the varieties of culture and environment. Rather than recruit professional film and television production crews, CESPAC selected persons with knowledge of and commitment to *campesino* development, and taught these people to use the video. The training lasted for four to five months and trainees learnt while producing video courses for *campesinos*. It was a comprehensive training programme that included the use of video equipment, routine maintenance of delicate equipment in difficult field conditions, topic research for programmes with the local communities, scripting and editing, and testing and evaluating programmes.[3]

The video production unit was small (two or three persons) and performed all production functions—from topic research to testing and evaluation. Non-specialisation was an advantage as each team member had a full understanding of the entire process and their training as evaluators ensured that audience reactions were fed back into the production process immediately. The production units organised screenings of training courses which were scheduled according to seasonal convenience and included suggestions for practical work to be undertaken by the *campesinos*. CESPAC believed strongly that the portable video production technology was very effective with the illiterate or semi-literate *campesinos* because of the medium's ability to show processes systematically and with adequate emphasis on details. The training courses included accountancy, animal health, maintaining farm machinery, dairying, alpacas farming and diagnostic courses focusing on problems of an area in order to stimulate discussion amongst farmers of possible solutions. Topic research ensured that the courses were locally relevant and the low cost and flexibility of the technology enabled the regional units of CESPAC to produce courses for specific groups. This would not have been possible if CESPAC had used the national television network and more expensive broadcast equipment. Most of the programmes were based on field recordings and the video technology speeded up the production process enabling the CESPAC team of 30 people in four production centres to produce 25 training courses with over 250 separate programmes of 10–20 minute duration in two years. Manuel Calvelo Rios best described the approach of the CESPAC project when he said: 'You ask a peasant, "Has it been well done?" The peasant will tell you, "Yes, we understand it." So, you leave it. It has not been well done? Then you play back, erase and record. This gives one enormous working advantage in production.'[4]

KHEDA RURAL TELEVISION

Following the SITE (1975–76) in India, the Indian Space Research Organisation (ISRO) established a rural television project in Kheda district of Gujarat near Ahmedabad, where the Space Applications Centre (SAC) was located. Four hundred direct reception TV sets were installed in the district's 350 villages and programmes for telecast were produced locally at SAC, Ahmedabad. The experiment aimed to improve upon the limitations of SITE and examine whether locally produced materials and use of the low-cost portable half-inch video portapak to focus on problems and issues raised by the local community could sustain the interest of the audience and bring about some change (Basrai 1976).

The Kheda team from SAC consciously chose to focus on the structural problems faced by the rural poor and used television to 'awaken' the audience in the Freirian sense. The programmes attempted to promote self-reliance among the community by showing that change was possible through optimal use of local resources without depending on external agencies. These awareness generation programmes were supported with instructional programmes on agriculture, animal husbandry, health and family planning, functional literacy and other developmental issues. The Kheda Rural Television also improved horizontal communication among the rural communities and initiated a dialogue between the people and the decision-makers in government through television. The prime target audience of Kheda Rural Television, according to the programme managers, were the rural poor who needed the 'catalytic input that would help them to help themselves'.

The production unit for Kheda Rural TV comprised a producer, a social science researcher and a scriptwriter working as a team and seeking help from a subject specialist when the need arose. This team mode of production was vital for the success of the Kheda experiment. Usually, in broadcasting organisations, the researcher has no contact with the producer and there is little field data available to the producer regarding the socio–economic profile of the audience, their language and cultural specificities or their felt needs. The scriptwriter of the Kheda team was consequently better informed and the presence of a researcher in the team also provided opportunities for pretesting formats and storylines. SITE had earlier shown that programmes produced with this kind of field research had better acceptance by the audience.

Production methods followed a pattern of maximum interaction with the villagers of Kheda (Karnik 1976; Karnik and Bhatia 1985). An idea for a series was discussed by the production team and developed further through formative research (audience profile and felt needs assessment). Further research and pretesting was done using the convenient portable video equipment. Based on this research, scripts were finalised and pretested with the villagers for comprehension and acceptance of the format. This ensured that communication gaps between the producers and audiences were minimised. Topic research ensured that the programme content was relevant and appropriate for the audience. Prototypes were pretested to make sure that the villagers would understand and like the programmes, which were modified—if necessary—based on the pretesting results. After the series was produced and telecast, the researcher gathered feedback for use in future programmes. Sometimes, producers viewed programmes with the village audience in order to get a first-hand experience of audience reactions.

Kheda Rural Television came close to the villagers and producers sometimes had to tread carefully on issues that were politically sensitive. Working for the rural poor and presenting their point of view on economic issues of oppression and exploitation did not please the powerful in the village and political pressures were brought to bear on SAC. Being a unit of the Central government helped somewhat, but even then there was a danger of exposing the vulnerable poor to further oppression for having spoken out before the camera. In an attempt to protect the poor villagers, producers came up with a combination of the documentary and drama formats, that enabled them to raise issues in a documentary programme and at the same time narrate stories of the exploitation of local villagers in a dramatised form with no reference to real people but set within the same cultural context. The message would thus go home but the vulnerable poor, who had begun to trust the production team, would be protected from further exploitation. This kind of 'brinkmanship' cultivated by the Kheda team was especially used in programmes that dealt with Scheduled Caste (SC) communities and their attempts to organise themselves and resist the oppression by the upper castes.

The portable video equipment gave the producers enormous flexibility. It could be slung on the shoulder and carried to the villages on a motorcycle. A two-member team could go out and do a recording. Tapes were erasable and therefore did not restrict the producer and village discussions could be recorded at length and edited later. Shot material could be played back immediately to check on quality and also shown to the interviewees and their comments noted. Villagers themselves could be encouraged to suggest programmes and direct the recordings. Playback of materials could be organised in the village to establish credibility and trust. Technically, the broadcast engineers found that the quality of the half-inch video was as good as the 16mm film televised and received via satellite relay on a TV set.

On a limited scale, and under special conditions of support from the highly powerful Department of Space, Kheda Rural Television successfully demonstrated that rural people could participate in the production process of television programming. Such local programming was of great interest to local communities and also gave 'voice' to the poor and the oppressed. It bridged the communication gap between the people and the authorities as information flowed back to the administrators and lapses and failures could be corrected quickly. Kheda Rural Television established itself as a community TV channel and the use of portable video equipment and the producer–researcher–scriptwriter team ensured that the programmes were grounded in the reality of the audience and appealed to them.

ALTERNATIVE COMMUNICATION TOOLS: EXPERIENCE IN THE VOLUNTARY (NGO) SECTOR

The radio in India was dominated by the government-controlled All India Radio, while the press was controlled by large business houses. Fortunately, the freedom of the press at least permitted small newspapers and periodicals to be published by committed and professional journalists, trade unions and political activists with limited resources. However, their reach was limited due to lack of advertising support. Broadcasting, on the other hand, was entirely in the control of the government and private participation was not allowed until well into the 1990s. Documentary film production was also controlled by the government through Films Division, which had the rights to distribute their films to all cinema halls that had to compulsorily screen one documentary or newsreel before the main feature film.

Until the late 1960s, production technology was restricted to 35mm film production equipment and broadcast quality full-track audio tape recorders. Independent filmmaking was consequently unaffordable. The few independent filmmakers that were there had to find corporate sponsors or be commissioned by Films Division. Television was only in Delhi and mostly studio based. Black-and-white silent footage of news coverage and human interest or development stories were accepted on a limited scale. It was a bleak scenario for making and distributing audiovisual materials with an independent perspective and covering development issues.

Technology was changing with 16mm and Super 8 film becoming easier to work with and meeting editing requirements and quality standards. Audio cassette recorders and low-priced microphones made recording audio programmes more easily possible. Commercial broadcasting (Vividh Bharati) had introduced many commercial studios where programmes could be produced as well. In a limited way, the tape–slide presentation and an occasional 16mm film were the alternatives to the mainstream images churned out by the government media or the entertainment film industry. The portable half-inch video recorder along with a black-and-white video camera introduced by Sony Corporation in the early 1970s came as a new entrant and competitor to the low-gauge (16mm and Super 8) film technology for industrial, educational and home use.

It was against this background that several institutions started using the new communication tools for creating alternative images and perspectives. Chitrabani in Calcutta, Xavier Institute of Communication Arts in Bombay, Centre for Development Communication in Hyderabad and Centre for Development of Instructional Technology (CENDIT) in New Delhi were some of the important nongovernmental agencies and institutions that came up in the early 1970s. Promoting the use of audiovisual media in education, training and development, these agencies produced slide–tape presentations, films and video programmes that supported the face-to-face communication work of voluntary agencies engaged in social work and rural development. Since training opportunities in audiovisual media were restricted to the single National Film Institute (later renamed Film and Television Institute of India), these institutions also trained interested individuals in audiovisual media production techniques using low-cost production tools—simple photography, audio recording, slide-tape and video production. Teachers, extension educators and development workers could be easily trained to script their educational materials as per their own requirements and, if possible, produce the programmes themselves.

Some of the early experience of recording village people expressing themselves using portable video equipment was demonstrated to AIR and ISRO at the time SITE was being planned. Several people argued that SITE could be used 'to give the community back to itself' by using television as a dialogical tool and social mirror.[5] However, the instructional nature of SITE and the multifaceted research design and programme management could not be changed at that stage. It was only after SITE and the learnings from the limitations of the professional broadcaster's transmission/persuasion paradigm that ISRO initiated the Kheda Rural Television project as a limited experiment in local programming using low-cost video equipment (the half-inch video portapak) and validated its use as a programme origination format for Indian television. Ironically, Doordarshan did not officially accept it (till the introduction of colour television in 1982) and continued with studio-based programming or field recordings on 16mm film.

If technology was making it easier for people to make programmes, organisations like CENDIT were facilitating voluntary agencies to make these programmes. International agencies like FAO, WHO, UNESCO and UNICEF were keen to capture the reality on the ground and promoted the participation of voluntary agencies as catalysts or change agents facilitating the development process. Audiovisual documentation of the work of such voluntary agencies on film and video gained importance. Trade unions and activists engaged in people's struggles for rights and entitlements used the new communication tools to document their struggles and enlist support from academics and the media. Informal distribution networks were created for limited circulation of such materials. Video captured images with an immediacy that was unknown till then. The medium was demystified and credible.[6] As audiences saw themselves in their own environment reflected on the screen immediately after recording, it became possible for them to believe that the 'screen tells the story of others like themselves, real and suffering, in other places and situations'. The rough nature of the field recordings that lacked the 'polished refinement of art' helped in exploring the nature of the problems faced by the people in a relationship of trust. Since the equipment was easy to handle, the involvement of development workers and village people in the production became possible and video became a more relevant and powerful aid in their work. Articulations of ordinary villagers and their comments on the inequity of the state's development agenda were unsettling for many of the development planners and administrators. The field experience of many fieldworkers was being brought back to the urban centres of planning and policy making. Video became very effective in crystallising issues and opening communication channels both within the community and between the community and the outside world.

NOTES

1. Colin Low was with the National Film Board of Canada (NFBC) and executed the Fogo Island Project of NFBC (see Low 1976).
2. TY 16 (16 years after the independence of Tanzania) video project was supported by FAO and a description and record of the experience is available in Schultz (1973).
3. The training in video production that CESPAC developed was very effective and useful for training the non-specialist. CESPAC developed a whole Methodology of Audiovisual Pedagogic Production (see UNDP/CESPAC/FAO 1985).
4. In personal conversation with Manuel Calvelo Rios when he visited India to participate in workshops organised by CENDIT for agricultural extension workers (1985) and development activists (1980).
5. The UN Panel of Experts meeting on SITE in New Delhi in November 1972 was a forum where many alternative ideas regarding choice of technology (Super 8mm and half-inch video, etc.) and alternative applications for SITE were presented. AIR also organised a national seminar on 'Software Objectives for Indian Television' (1973) where the social and development purpose of television was discussed.
6. Use of alternative images and convenience of video is described in an early paper by Akhila Ghosh (1986).

Networks and Movements

In developed countries people were being alerted to the devastation of the environment caused by rapid industrialisation and indiscriminate exploitation of natural resources. The first UN Conference on the Human Environment was held in Stockholm in 1972 and rapidly thereafter concern over air and water pollution, deforestation and soil erosion became widespread. The UN World Population Conference in Bucharest in 1974 was followed by the first UN World Conference on Women in Mexico City in 1975. Both these conferences focused on other important issues of global concern. Increasingly, it was recognised that the media had an important role in redefining development priorities. The focus shifted to building pressure on governments through citizen's groups and associations of individuals (broadly referred to as non-governmental organisations or NGOs) and other civil society initiatives. The oil crisis of 1973 and the sudden rise in oil prices made everybody realise that there were limits to growth as the earth's non-renewable resources are finite.

In India, the women's movement and the environment movement that began in the 1970s were two very significant initiatives that altered the way in which development was defined. For the first time, women's role in economic activity and their contribution in the home began to be measured. Domestic violence, unreasonable demand for dowry leading to 'dowry deaths' caused by unbearable oppression of the victim (including dousing them in kerosene and setting them aflame) came to be reported in the media and drew the attention of the public. The women's groups that sprang up in cities began fighting for equal treatment and demanding stern action against eve teasing, sexual harassment and dowry. The portrayal of women in cinema alternately as the archetype of virtue and submission or the corrupted modern, Westernised 'vamp' began to be questioned by them.

While much of the attention on women's issues may have been initiated in the cities, the quiet suffering of rural women also came to the foreground. Their poor health caused by malnourishment and discrimination against the girl child, repeated childbirth as well as the back-breaking labour in the fields and at home was a 'double burden' that cried out for justice and relief.

Women's groups adopted new forms of communication to create awareness against injustice and oppression, denial of basic rights, torture and domestic violence. Street-corner plays followed by discussions with the audience were a popular form in towns and cities, particularly against dowry. Audiences were moved by the narration of victims portrayed in the plays and the discussions that followed often led to the formation of neighbourhood vigilance groups and establishment of centres where women in distress could seek shelter, support and guidance.

In rural areas the mode of communication adopted was one that was more appropriate for rural women. The *Mahila Mela*—a gathering of rural women from different parts of India—organised at the Social Work and Research Centre (SWRC), Tilonia, in 1985 was one such novel initiative. Several voluntary agencies in Rajasthan came together to organise this week-long event in which about 900 women from various parts of India participated as individuals and as representatives of their organisations. For many of them it was their first move out of their home environment. The unique quality of the event was the easy pace at which the women related to one another, overcoming barriers of language through gestures, singing, dancing, painting, acting and sharing the work and joy of being together in an uninhibited manner.[1] Rural women were exposed to women from other parts of India and learnt a great deal from one another. They did not suffer from the inhibitions of urban women and spoke in their own language, leaving it to others to translate, if they could. Taking their cue from the rural women, the urban, middle-class women activists learnt to shed their inhibitions and understand the problems of the former a little more keenly.

Participating in discussions and expressing their views before a large gathering with their voices amplified by the microphone gave these women self-confidence and a feeling of collective strength. The spontaneous nature of the event allowed participation by anybody and otherwise taboo subjects like rape were discussed. The father of a 12-year-old girl found the courage to take the microphone and narrate to the entire gathering the rape of his daughter and the subsequent inaction by the police. Others listened with rapt attention and the gathering passed a resolution following which a black-armband silent procession marched through the village of the rapist and surrounded the concerned police station demanding the reopening of the case. This kind of spontaneous self-expression and coming together became a powerful tool of communication that encouraged a feeling of solidarity and collective strength in the women's movement.

In a different way, there were other experiences of producing audiovisual materials that challenged existing myths and misconceptions about women's status and position in society. The traditional male–female (dominant–submissive) role was questioned through these materials. A series of five audiovisual programmes, *Kahani Nahanachi*, produced by a group of women (Astha, Bombay) in 1981–82 and *Dohra Bojh*, a video programme documenting a day in the life of a Harijan woman produced by CENDIT and Action India, New Delhi, in 1981, are two examples of this kind of early exploration.

Kahani Nahanachi was produced through extensive discussion with women residents of urban slums in Bombay and dealt with issues of women's reproductive health—menstruation, pregnancy and childbirth, and sexuality. The premise was that women needed practical information that would be useful to them. At the same time, the producers wanted to raise questions about the traditional role of women as 'wife' and 'mother'. A simple explanation of X and Y chromosomes and the discovery that the sex of the child was in no way determined by them was a great relief to the women who used to suffer much guilt and despair at being blamed for something that they were not responsible for: the birth of a girl child. The form of presentation in the tape-slide mode attempted to break new ground. The use of an off-screen commentator was abandoned and the entire presentation was done through a semi-musical dialogue between two characters, Ajibai (grandmother) and Taruni (young, modern woman). While this format was attractive and appealed to the audience, the resolution of the conflict in views and acceptance of the modern viewpoint seemed to be too easy. Based on this feedback after the screening of the first programme, the producers modified the format and changed it to a first-person narration of a character easily identifiable by the audience as one of them. She discussed problems of pregnancy and childbirth through her own experience and those of her friends and gave information without offering easy solutions; instead, she alerted the audience to aspects of health norms which were always conditioned by other aspects of life circumstance. These audiovisual materials were translated by other women's groups and used extensively and the evaluation report by Anjali Monteiro[2] became a starting point for discussion about presenting alternative images of women and discussing issues from a feminist perspective.

In Delhi, CENDIT worked with Action India, a social organisation, on a video programme entitled *Dohra Bojh* (Double Burden). The programme centred on the daily life of two women (mother and daughter) in a Harijan family. Apart from the physical activity of cooking and cleaning and working in the fields and back to cooking in the evening, the woman bore the emotional anxiety of a son truant from school and a married daughter rejected by her husband and ill with tuberculosis. *Dohra Bojh* portrayed the enormous physical and emotional burden placed on the woman at home, the lack of her place in society, and the emptiness and absence of any kind of reward. Further subjugation, rather, was the norm. Before showing the video programme in the community the family was shown the edited material in confidence to ensure that their privacy had not been violated. The family and the women felt strongly that it was important to show the film as it told their true story and others suffered similar fates. When *Dohra Bojh* was shown around it acted as a discussion-starter, provoking and eliciting reactions for further discussion. The conventional portrayal of women in the media did not show any alternative roles for women in society nor was it realistic about their feelings and capacity to alter themselves. The media generally supported the attitude of others, especially men, towards them by predominantly presenting the image of a helpless woman. *Kahani Nahanachi* and *Dohra Bojh* for the first time began questioning this kind of portrayal and presented alternative perspectives that reflected the concerns of ordinary women and also depicted their strength and resilience in the face of adverse circumstances.

The women's movement benefited from the international Women's Liberation Movement and the feminist agenda and debates in post-industrial developed societies. Even if the economic and social circumstances were different, the fundamental issue of male dominance or patriarchy and consequent injustice remained the same and the women's movement in India created an initial wave of awareness and enthusiasm that brought women's issues to the centre of the development debate in this country.

The protest against the multipurpose irrigation and power project in the Silent Valley in Kerala and the subsequent abandoning of the project could be regarded as the first significant achievement of the environment movement in India. A unique feature of the protest was the widespread local support for the environmental movement in Kerala which included India's most well-educated population with strong political leanings towards the left. In fact, the Kerala Shastra Sahitya Parishad (KSSP), a group of scientists, academics and activists engaged in popularising science among the people of Kerala, was initially hesitant in supporting this movement as it appeared to be against industrialisation (or modernisation) and economic development. However, the arguments put forward by the environmentalists computing both the damage to the flora and fauna that was particular to the pristine equatorial rainforest of Kerala as well as the long-term ecological destruction found international support and the movement grew in strength through the awareness campaign carried out in the towns and villages of Kerala through *kalajathas*—mobile theatre performing troupes that integrated visual display through posters and discussions in small groups after the performance. The media picked up the popular resistance to the Silent Valley and a countrywide protest gathered momentum and put pressure on the Central government to finally abandon the project. The success of the KSSP *kalajatha*[3] effort in Kerala and particularly the Silent Valley movement led to the spread of this mode of communication that was more 'people-centred' and interactive to other parts of India. When the Oleum gas leak disaster occurred in the Union Carbide plant in Bhopal, spontaneous protests were launched but the culmination of that effort in educating people on the harmful environmental consequences of industrialisation came with the countrywide *jatha* that was organised by a network of agencies working towards popularising science among people. Performing troupes of activists travelled from the four corners of India giving performances in towns and cities on the way, showing the horror of the Bhopal gas tragedy and documenting the other environmental consequences of industrialisation in terms of air and water pollution, displacement and deforestation. In 1986, one year after the accident in the Union Carbide, all the groups came together in Bhopal—an event which received wide media attention and demonstrated the strength of networking and effectiveness of popular modes of communication.

The other significant protest movement on an environmental issue has been the Narmada Bachao Andolan (NBA) that has been ongoing for over 15 years. Again, the people likely to be displaced by the large dams are at the centre of the struggle, indicating the strength of the 'bottom-up' mode of communication. This struggle has been bitter and polarised between those in Gujarat who are to receive short-term benefits from the large dams and those who will lose their homestead and land with little or no compensation. The quality of research and lobbying through the media and the courts, as well as the popular resistance on the ground generated through marches and rallies, songs and sit-ins, has been unique in its sustaining power. Even if the project has not been abandoned completely, the NBA struggle has clearly shown that the people cannot be taken for granted and that the knowledge and expertise of urban planners is often flawed and questionable.

In a different vein, another major initiative of the environment movement was crystallised in the form of the *First Citizen's Report on the State of India's Environment* published by the Centre for Science and Environment (CSE) in 1981. This was a demonstration of the effective networking of individuals, agencies and institutions documenting the vast scale of devastation and destruction wrought by the mega-scale development projects initiated by the government without any calculation of the loss in environmental (natural) wealth. Nor had the planners computed the effects on the culture and lifestyle of people who lived in harmony with their environment in the hills and forests, deserts and coastal plains of India. The *Citizen's Report* was unique because it provided an insight into the myriad efforts of individuals, small groups and agencies in different parts of the country trying to find alternative paths to development that would conserve rather than destroy the environment. The Chipko movement in the Garhwal Himalayas where the hill women clung to trees to stop their felling for commercial profit while the springs ran dry and the loosened soil led to landslides, or the people in Sukhomajri village near Chandigarh who worked together with a government research institute to restrict wanton grazing and thus leading to afforestation of wastelands and better irrigation of their fields are some examples of micro-level experiences. Such initiatives brought home the fact that people cannot be taken for granted and that their wisdom and capacity for innovation should be utilised in development planning that is, in the end, meant for them.

The women's movement and the environment movement have been highlighted to illustrate the innovative means of communication that were adopted for reaching out to the people. Some of the local forms used drew inspiration from the days of the freedom struggle while others borrowed from the 'agitprop' street-corner plays of the socialist/communist movement and yet others innovated on the rich tradition of folk theatre and other performing arts prevalent in different parts of the country.

RESPONSE FROM THE STATE

It should be mentioned that the 1970s was a period of great political stirrings throughout the country. The decade began with the liberation of Bangladesh in 1971 and the massive mandate given to Indira Gandhi in the general elections soon afterwards in 1972. Her populist programme of *garibi hatao* (remove poverty) did not deliver on promises and political ferment reached its peak under the charismatic leadership of Jaya Prakash Narayan in 1974. The imposition of Emergency and consequent press censorship in June 1975 alienated the people further. When the country went for general elections in 1977, the Congress party failed to get a majority in Parliament for the first time since 1952 and the first non-Congress Janata Party government was formed under the leadership of Morarji Desai.

Much hope was placed on the new government but intra-party feuds soon came to the surface causing the government to fall and Indira Gandhi returned to power in 1979. Perhaps chastened by her defeat in 1977, the new Congress government tried to make a new beginning with greater concern for a more equitable and human development paradigm. However, the political turmoil and divisive forces operating in different parts of the country, particularly in Punjab and the North-east, continued to preoccupy the Central government. It led, finally, to the military Operation Bluestar mounted to flush out the Khalistan militants and the subsequent assassination of Indira Gandhi in 1984.

The turmoil of this period coincided with the upsurge of political and popular protests from the people against the increasingly authoritarian role of the government. In particular, they were questioning the credibility of the state, and there was widespread disillusionment with political parties in general. Spearheaded by voluntary agencies and what Rajni Kothari termed 'non-party political forces' (NPPF), a movement for creating a political space for participation of people in raising issues of human rights and development policy was gaining ground. Women's rights (and the rights of other ethnic and social minorities) and the environment movement represented the coming together of such groups under a broad agenda.

The response from the Rajiv Gandhi government after 1984 was to try and induct and accommodate voluntary agencies in the planning process itself, introduce major policy reforms in education with greater concern for women's equality through education, and shift the focus to basic education—Universalisation of Primary Education/ Universalisation of Elementary Education (UPE/UEE) and adult literacy. To meet the basic needs of people a new management approach towards development was introduced through the establishment of Technology Missions in the areas of drinking water supply in rural areas, childhood immunisation, adult literacy, production of oilseeds and rural telecommunications. The idea was to apply the science and technology expertise available in the country towards people-centred development problems. One of the key ideas of the Technology Missions was partnership and transparency. State governments, NGOs and voluntary agencies, science and technology research institutions and other academic bodies and the media were all involved in the missions which had a time-bound, goal-oriented approach towards the achievement of development objectives. People's participation was crucial to the implementation of programmes under these missions. Sam Pitroda, Adviser to the Prime Minister on the Technology Missions, observed: 'These programmes cannot be implemented by the Central government nor by the state government. It has to become a people's movement.'[4] The government therefore took an active interest and played an important role in communication and social mobilisation to generate demand for basic services.

Social mobilisation was defined as 'the process of bringing together all feasible and practical inter-sectoral social allies to raise people's awareness of and demand for a particular development programme, to assist in the delivery of resources and services and to strengthen community participation for sustainability and self-reliance' (McKee

1992). This inclusive definition attempted to shed the controlling authority of individual development departments of government and sought cooperation among them, inviting the participation of all sections of the community for a specific goal like literacy, childhood immunisation or family welfare. Usually, social mobilisation was regarded as a one-time intensive effort or a campaign to create a favourable and supportive climate for a particular development programme. It was expected that once the community (or beneficiaries) became aware and the demand for services increased, the service providers would respond swiftly to fulfil the unmet needs. However, the reality on the ground did not correspond to this expectation because the government functionary or service provider was not adequately trained to be responsive to demand generation. This led to disappointment and loss of momentum. However, the Technology Mission's approach towards social mobilisation, building on the positive experience of the Expanded/Universal Programme of Immunisation (EPI/UPI) initiated in collaboration with UNICEF, was a significant communication experience that we shall examine in greater detail in the next section (Section II) in connection with the National Literacy Mission (NLM).

There was some criticism of the Technology Missions' corporate management approach. There was also some apprehension regarding the induction of voluntary agencies in the development process through state patronage (thereby limiting their adversarial role in critiquing state policy and programmes). It was acknowledged that the government was moving towards more openness and inclusion of basic issues and concerns of the large majority of people in its development agenda. This was also necessary because of the need for consensus in a fractured polity divided by caste, class and gender as well as along regional, linguistic and religious lines. In the words of Rajni Kothari, 'the new conceptualisation that is emerging is built around the notion of rights and liberties of various sections of people ... rights that obtain from social, ethnic, ecological, gender and ethical mainsprings of a diverse and plural society'(Kothari 1989).

In Section II we examine the experience of state-led initiatives during the last decade (1990s) in the major areas of development concern—literacy and basic education, population and reproductive and sexual health, and facets of rural development and decentralised governance—where communication effort has played a significant role in altering the nature of discourse on development issues. We shall ground our analysis on a framework of communication for human development that builds on the experience of earlier decades after Independence and tries to include wider participatory and dialogical approaches.

NOTES

1. A film documenting the Tilonia *Mahila Mela* was produced by CENDIT (1986) and the report of the first workshop in 'South Asia on Women and Media in Development' (1986) held in New Delhi describes the *Mahila Mela* and many other media-based initiatives for alternative imaging of women in the South Asian region (see Kapoor and Kapoor 1986).
2. The report and the audiovisual programme was prepared for Astha, an audiovisual resource group working in Bombay.
3. A more detailed account of the *kalajatha* form of communication is given in Section II in the context of the Total Literacy Campaigns. The genesis of the *kalajatha* was in the 'people's science movement' in Kerala initiated by Kerala Shastriya Sahitya Parishad (KSSP). The all-India coverage through a series of *kalajatha*s finally assembling in Bhopal (1986), one year after the Oleum gas leak, was the first massive use of this form of alternative communication for creating awareness.
4. Observation made by Sam Pitroda in the film *Literacy?* (1988) produced by the Directorate of Adult Education, New Delhi, in connection with the launch of the National Literacy Mission (NLM).

Communication and Human Development: A New Approach

In the preceding chapters we reviewed the development paradigm pursued in India since Independence with the belief that mere information regarding the modernisation or development programmes would be sufficient to instigate change in the cultural habits and prevailing traditional practices. While some change has come about, it is the failure of the state to deliver on promises of a better standard of living and greater opportunities for employment and income that has been disappointing for the people.

It is true that the state has been beset with the problems of poverty and burgeoning population and the enormous task of building the physical infrastructure for achieving development objectives. It is equally true that the benefits of the development programmes have been more favourably skewed towards those with education and social advantage in spite of several legislations, reservations and other affirmative actions to end social discrimination and injustice.

By the beginning of the 1980s the growing middle class in urban areas, who had become economically powerful and vocal, demanded the removal of strictures on consumption. The introduction of colour television in 1982 and its commercial exploitation for revenue generation thereafter are perhaps the clearest illustration of this tendency. While progress with expansion of primary schools and primary health care facilities remained tardy, low-power transmitters to relay television signals/images from New Delhi all across the country spread at a rapid pace!

The state's professed commitment to eradication of poverty and illiteracy, improved health care facilities and better employment prospects was not matched by the performance on the ground. Equally, the role of the traditional (colonial) administrator as a dispenser of justice and 'friend of the poor' was being eroded by its own development policies and programmes that were destroying the economic, cultural and social base of large numbers of rural people and forest dwellers. The state's inability to deliver on its promises and its indifference to the miserable plight of the people created resentment and loss of credibility among the people. Hence, cynicism and apathy towards well-meaning government initiatives prevailed and there was no sense of belonging, ownership or participation by people and the community in development programmes. The common belief was that the benefits of such development programmes accrued to the rich and powerful—the politician, the bureaucrat and the contractor.

At the same time, there was a growing awareness of human rights and basic entitlements even amongst the unlettered and disadvantaged. The exercise of adult franchise in the periodic elections to state legislatures and the Indian parliament, as well as the spread of information through the media and the grassroots-level efforts of voluntary agencies and NGOs contributed towards the creation of this awareness. As we have seen earlier, the environment movement and the women's movement in particular, were effective instruments in bringing gender injustice and environmental concerns upfront in the development agenda. This would not have been possible had there not been a sufficient groundswell of popular support to these causes. The adversarial position adopted by activist groups in confronting state policy and development programmes found support from the judiciary and the political opposition for the government to reformulate its development agenda and enlist greater people's (community) participation.

This was in part also a result of the growing market penetration by and rising consumer awareness about a wide range of 'branded' consumer products—from soft drinks to washing powder and torch cells to tea-packets. Surely there was a lesson in this for the social sector? If marketing of consumer goods was possible even in situations of low purchasing power, was it not possible to instil motivation for demanding basic education and health care? If people came forward to demand services then, it was argued, the supply of services would have to follow. This was a move away from the earlier 'supply push' model that was not delivering quality or satisfaction. A concerted effort was made in this direction at least in one programme launched by the Government of India in the mid-1980s in collaboration with UNICEF—the Universal Programme of Immunisation (UPI) for child survival in the 0–5 age group all across the country.

A basic feature of the UPI was the task-oriented, management-driven approach with primacy given to the beneficiary. Targeting the mother in every communication to ensure that the child received the complete immunisation schedule, training the frontline worker, facilitating the logistics of supply and delivery of immunisation services and monitoring the process closely with community support were some of the key elements of this programme. The UPI (later known as the Expanded Programme of Immunisation—EPI) and the promotion of breastfeeding were two important child survival concerns of UNICEF worldwide. International experience from several countries, use of research data and professional agencies to design and implement a strategy for generating demand for services, and mobilising political and community support for the programme gave rise to a new understanding and approach to communication for development. This was more holistic, participatory and sensitive to the circumstances of those in need of the services.

The new approach to communication for development made a serious attempt to draw on the lessons from the earlier experiences in development communication. The most important learning had been that no development programme could succeed without political consensus and widespread community support. The government by itself could not realise the goals without societal acceptance of the programmes and active community participation to achieve the targets. Further, specific behaviours and practices, individual and collective, had to be influenced and changed through focused communication based on an understanding of prevalent practices and addressing the reasons for resistance to change. All this required a more professional, research-based approach to communication that included training the field staff with new communication skills based on an understanding of and respect for the beneficiary groups or persons whose behaviours needed to be changed. The communication process required periodic monitoring, assessment and fine-tuning to ensure behaviour change. In other words, communication was no longer a one-time effort to disseminate messages through different media channels but an engagement and an ongoing exchange between the government and the people to design and implement the development agenda.

Figure 5.1
Communication and Development Model
Source: McKee (1992).

Communication thus assumed a more important and critical role in the present context of wider democratic participation and economic liberalisation.

In this model[1] of communication for development (Figure 5.1) there are three elements that work together—advocacy, social mobilisation and programme communication that addresses specific behaviour change concerns. All three elements have to act in concert and require to be supported by research and monitoring and training of field personnel in communication skills and management of strategic communications.

Advocacy is the marshalling of facts and information into persuasive arguments for a particular development objective. This has to be addressed to the political leadership (at all levels), planners and academics, media and the lay public in order to arrive at a consensus regarding a particular development programme (e.g., routine immunisation, reproductive rights, decentralised governance, safe drinking water, etc.). Advocacy is also not a one-shot affair because the commitment of the political and social leadership has to be renewed till the problems are overcome and the development objectives are accomplished. While the audience for advocacy may not be required to modify their own behaviour in many cases, their attitude towards social discrimination, economic deprivation, equity and injustice greatly influence social values and norms that in turn affect (influence) individual and collective behaviour. Preference for a male offspring, lower status accorded to women, indifference to the exploitation of child labour are examples of prevalent social values where advocacy can heighten concern and commitment to bring about change.

Social mobilisation may be regarded as the endeavour to bring together inter-sectoral alliances and partnerships between private and public institutions and agencies to create awareness on a particular issue. Usually, it is an effort on a sufficiently large scale through coordinated communication activities that results in cohesive action. Social mobilisation is a means to generate demand for a development programme or a service (drinking water supply and rural sanitation, reproductive and child health care, primary education, etc.) and ensuring allocation of resources and delivery of good quality services that meet the needs of the people. The inter-sectoral alliances are necessary to encourage community participation and sustainability through self-reliance. In the Indian context, the Bharat Gyan Vigyan Jatha (BGVJ) for adult literacy (and primary education) in the early 1990s remains the best example of social mobilisation in terms of people's participation. The Pulse Polio Immunisation (PPI) since the 1990s has also been a good effort which has showed that effective implementation of National Immunisation Days was possible through inter-sectoral alliance and field-level coordination.

The third key element of the new approach to communication for development is *specific communication in support of development programmes* to effect behaviour change among particular groups. This is often also referred to as programme support communication (PSC) or communication for behaviour change. While these terms have slightly different connotations in usage by different agencies, the focus, for our purpose, is on the approach to effective communication to bring about change in behaviour. It is based on an understanding of the audiences that have to be addressed through communication, the messages (ideas, information and emotional appeal) to be exchanged, the media channels to be used and the expected outcome to be measured through well-defined indicators. This approach draws from the theory and practice of marketing products and services in the commercial business sector and applies it to the social sector. Sometimes it is referred to as social marketing, though that term has different connotations in different contexts as the term 'marketing' is broader than 'communication'.

The communication effort in this case begins with an understanding of the audience and their present behaviour. This leads to an analysis of their resistance to adopting new behaviour/practices promoted by a social development programme—washing hands before eating or sending children to school. Audiences are placed on a continuum of behaviour change—exposure to new ideas, repeated exposure, discussion among peers and community, willingness to try a new idea (behaviour), finding social support for new behaviour, trying a new behaviour, learning experience from new behaviour, and finally stabilising on new behaviour.

This schematic diagram (Figure 5.2) shows the manner in which behaviour change takes place. At a societal level, the new behaviours in particular groups stabilise only when social norms change and there are enabling factors facilitating that process. This is the BASNEF (Behaviour, Attitudes, Social Norms and Enabling Factors) model (Figure 5.3) that postulates that behavioural intention is based on an individual's beliefs and social norms.[2] These change only when social norms change and individual beliefs are influenced through communication. Also, there have to be enabling factors in the environment—primary school nearby, easy access to a safe water source— facilitating that process.

Creating a *behavioural intention* requires understanding and dealing with the *values* that individuals place on a behaviour in personal terms, as well as in terms of social support/opposition to their actions. Again, an intention in itself cannot lead to a desired behaviour unless *enabling factors* are already in position: hand pumps at convenient locations *and* regularly maintained, low-cost latrines available at an affordable price, etc.

The components of a communication strategy are (*a*) to segment the audience and identify specific behaviour changes, (*b*) develop message concepts, and (*c*) plan a mix of media based on patterns of media use. The process of developing a strategy and plan for a programme initiative on communication for behaviour change begins with a problem analysis and identifying desired behaviour (or more comprehensively, Knowledge, Attitudes and Practice [KAP]) for each problem/issue, listing the existing KAP/behaviour of the target groups whose behaviours need to be changed (including influencers and service providers), developing message statements that need to be

COMMUNICATION AND HUMAN DEVELOPMENT

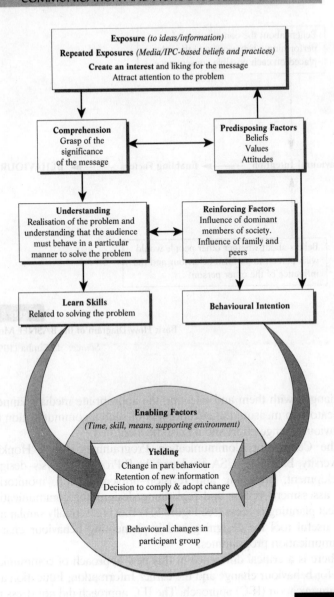

Figure 5.2
Behaviour Change Communication Process
Source: UNICEF.

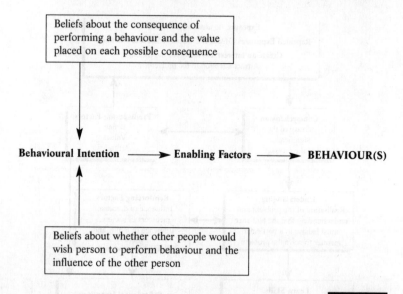

Figure 5.3
Basic Flow-Diagram of the BASNEF Model
Source: da Cunha (1993).

exchanged with them and selecting the appropriate media channels. Indicators to measure the effectiveness of such a communication for behaviour change effort should also be identified.

The Centre for Communication Programmes, Johns Hopkins University, Baltimore, USA developed the P-Process (analysis–design–development, pretesting and revision–implementation, monitoring and assessment–review and replanning/design) for communication project planning process (see Figure 5.4) that is essentially similar and is a useful tool for designing and implementing behaviour change communication programmes.

There is a critical difference in this new approach of communication for behaviour change and the earlier Information, Education and Communication (IEC) approach. The IEC approach did not stress the final outcome of behaviour change and relied on the dissemination of information (message), educational process and communication delivery system to be sufficient to bring about change in behaviour that would be reflected in an increase in demand for goods and

COMMUNICATION AND HUMAN DEVELOPMENT

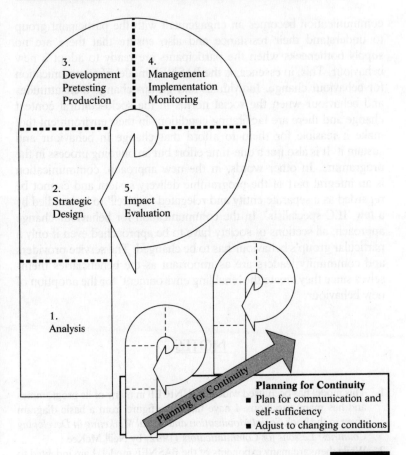

Figure 5.4
Communication Planning (P-Process)
Source: Johns Hopkins University.

services. Hence, the task of those responsible for IEC ended with the 'message' delivery or 'creating awareness'. A feedback mechanism may have been used to improve the design and content of the message but the role of IEC ended with the delivery of the message.

In the new approach of communication for behaviour change the *final objective is behaviour change* and its sustenance over time. There has to be an unrelenting effort by communicators to achieve this goal and, to that extent, every person involved in the programme has to be involved in the communication process. This means that communication becomes an engagement with the participant group to understand their resistance and also ensure that there are no supply bottlenecks when the participants are ready to adopt a new behaviour. This, in essence, is the BASNEF model of communication for behaviour change. Individuals and groups change their attitudes and behaviour when the social norms in the socio-cultural context change and there are facilitating conditions in their environment that make it feasible for them to afford that change in behaviour and sustain it. It is also not a one-time effort but an ongoing process in the programme. In other words, in the new approach, communication is an integral part of the programme delivery system and cannot be regarded as a separate entity and relegated to a 'cell' to be handled by a few 'IEC specialists'. In the communication for behaviour change approach, all sections of society have to be approached even if only a particular group's behaviour has to be changed. The service providers and community leaders are as important as the beneficiaries themselves since they create the 'enabling environment' for the adoption of new behaviour.

NOTES

1. This model has been used widely by UNICEF in many of its programmes and has many variations. I have taken the figure from a basic diagram given in the book *Social Mobilization and Social Marketing in Developing Countries: Lessons for Communicators* (1992) by Neill McKee.
2. While there are many exponents of the BASNEF model, I am indebted to Gerson da Cunha for his exposition on several occasions. His paper 'Safeguarding Youth from AIDS' (1993), describing the project and movement in Uganda and prepared for UNICEF and the Government of Uganda, is an appropriate and useful reference.

Section II

RECENT EXPERIENCES IN INDIA

Use of Communication for Literacy and Empowerment

Soon after becoming prime minister, Rajiv Gandhi gave a call to lead India into the twenty-first century based on the induction of new technology, professional management and human resource development. He initiated a review of the existing education policy and urged that science and technology be used to address the major problems of illiteracy, infant mortality, lack of drinking water and other social ills. Together, these measures presented a refreshing attitude towards human development and captured the imagination of the middle class in India. The series of discussions and meetings that preceded the formulation of the New Policy on Education[1] allowed wide participation by parents, teachers, planners, researchers, social workers and academics in the process of its formulation. Public opinion was built on the need for greater investments in basic education and correction of the gender discrimination in educational opportunities.

Eradication of adult illiteracy was one among the five Technology Missions established at that time to address basic problems. Operation Blackboard (OB), which aimed to provide every child easy access to primary education and set down norms of basic facilities that were to be available in all primary schools, and the National Literacy Mission (NLM), which aimed to make 80 million adults literate by 1995, were the two countrywide and significant programmes launched with great enthusiasm.

Beginning with the discussions on the New Policy on Education in 1986, there has been a consistent shift in public perception regarding the need for greater emphasis on basic education, particularly for girls. The launch of the National Literacy Mission (NLM) in 1988 as a societal mission to make 80 million adults in the age group 15–35 years literate by 1995 was a major initiative that was adequately picked up by the media. When Ernakulam district became fully literate through the literacy campaign initiated in the district with the active cooperation of the district administration, educational institutions and a prominent NGO—Kerala Shastra Sahitya Parishad (KSSP), it was a significant achievement. Publicity and coverage in the media helped in the adoption of the campaign mode as the strategy for NLM. Kerala achieving fully literate status soon afterwards, in 1990, which was also the International Literacy Year (ILY), made news and brought literacy to front-page headlines of newspapers. In that sense, the Jomtien Meeting[2] and the Education for All (EFA) declaration came when India was poised for a shift in priority and public opinion reflected in the editorial coverage and political manifestos endorsed the commitment towards the goal of Education for All by 2000.

It was the convergence of the concern for healthy development and basic education of the girl child (1990 was also the SAARC Year for the Girl Child), the upsurge of the voluntary spirit manifest in the literacy efforts initiated in several districts and the early success of the literacy campaigns in Kerala and pockets of other states, supported by positive reporting and coverage in the media, that ensured that successive governments and political parties of different hues gave due importance to basic education. No doubt international pressure was mounting and external funding was being made available for several basic education projects in the different states. Through the literacy campaigns a greater demand for primary education was becoming evident as parents showed interest in getting their children, including daughters, educated.

While the media played a role in the issues and concerns for basic education thereby creating an ambience in favour of greater investments and innovative activities, this role was largely restricted to the educated middle class and the ruling elite who were the key players in policy planning and implementation. However, while advocacy and opinion building is one critical area where the media have a role, another equally important area is that of demand generation.

Very often, the beneficiaries of a government welfare programme are unaware of the facilities and benefits on offer. The credibility of the government programme, the high-handed and officious attitude of government functionaries, the poor track record of teachers and educational administrators to care for the poor or socially disadvantaged, are all factors that have contributed to the alienation of the poor from government programmes. Local communities also despaired at the petty corruption and unresponsive manner of functioning of government departments. In the face of such negative conditions, any new initiative for Education for All (EFA) had to establish its credentials and demonstrate its sincerity of purpose.

Unless the intended beneficiaries came forward and the local communities ensured that the facilities set up by the government function well, the very purpose of the initiative would be defeated. Parents had to be persuaded to send their daughters to school regularly and not withdraw them at the slightest pretext. At the same time, schools had to be conveniently located and provided with at least the minimal facilities of toilets and drinking water, and social discrimination and humiliation of the poor and disadvantaged in schools had to be ended so that the school could become an inviting environment where children felt secure and happy. Planning facilities and providing teachers were not the solution. Parents had to be motivated and assured of the benefit of sending their children to school.

Here, communication media played a role in demand generation by providing basic information and creating awareness and also demonstrating the goodwill and sincerity of the administration. While the mass media (radio, TV, newspapers and cinema) had a role, it was the local media like folk plays and performances, exhibitions and *mela*s and community meetings that had a greater persuasive capacity. What was true for enrolling children in schools was also crucial for drawing adults to the literacy classes. Children by and large were expected to go to school to study. It required a high degree of learner motivation to join an adult literacy class. Social mobilisation, as amply demonstrated in the literacy campaigns, was the key to the success of the programme.

Broadly speaking, communication media contributed in motivating volunteers and functionaries as well as encouraging learners to participate in the programme. Whether it was the folk media and cultural *jatha*s,[3] as used during the Bharat Gyan Vigyan Jatha (BGVJ) in 1990 and 1992, or again in the Samata women's *jatha* in 1993 or in the cultural *jatha*s in the Total Literacy Campaign (TLC) districts, or the carefully designed radio and TV spots or the literacy song *Padhna Likhna Seekho* (learn to read and write) based on a poem/song by Bertolt Brecht, the objective was the same: make an emotional appeal to the viewer to identify with the cause of eradication of illiteracy or, in the case of learners, realise the necessity of becoming literate and so desire to become literate.

NLM—COMMUNICATION STRATEGY DEVELOPMENT

When the National Literacy Mission (NLM) was launched in May 1988 as one of the Technology Missions, it was clear that its objectives could not be achieved unless it became a 'people's movement'. NLM required the participation of people—the educated as volunteers to make adults literate, and the unlettered to come forward as learners and become literate.

Literacy itself was perceived as a communication skill and learning to read and write enabled learners to better understand the world and express ideas and share experiences more widely. The objectives of NLM[4] went beyond learning the traditional three 'Rs' (reading, writing and arithmetic). It implied and included:

- Achieving self-reliance in literacy and numeracy.
- Becoming aware of the causes of their deprivation and moving towards amelioration of their condition through organisation and participation in the process of development.
- Acquiring skills to improve their economic status and general well being.
- Imbibing the values of national integration, conservation of the environment, women's equality, observance of small family norm, etc.

It was an uphill task because adult literacy was not regarded as a worthwhile investment or pursuit. While there was a general consensus regarding primary education and Universalisation of Elementary Education (UEE) was seen as an urgent necessity, adult literacy was not taken seriously at the time when NLM was launched. However, the New Policy on Education articulated the need to engage women as learners. The agency of women in the new development paradigm was also beginning to be recognised by planners and academics.[5] Women's literacy, therefore, was conceded some priority though economic independence and forming women's collectives/groups or organisations were given greater importance.

The NLM document laid down certain targets to be achieved during the life of the mission. Eighty million adults in the age group of 15–35 years were to be made literate by 1995 (later increased to 100 million by 1997—the end of the Eighth Plan period). In the entire programme, the focus was primarily on poor women, both urban and rural, and persons from Scheduled Castes (SCs) and Scheduled Tribes (STs). It was realised in the NLM document that the government alone could not accomplish a task of this magnitude. Also, substantial improvement in the efficiency and quality of the service being provided at that time was required. Therefore, the idea was to encourage people's participation in programme implementation and improve the quality of performance of all functionaries. Since the delivery system needed to be upgraded it was necessary to reduce wastage, that is, the gap between enrolment and completion of the literacy course had to be reduced and relapse into illiteracy prevented. In order to attract the adult learner, the total environment of the literacy class (including literacy materials, pre-literacy motivation and the post-literacy and continuing education programmes) needed improvement. Apart from the necessary training impetus for functionaries, it was also important to improve internal communication within the existing system and structure of NLM so that the flow of information from the field to the district and state capital as well as the NLM headquarters in New Delhi (and the system's response) was quickened.

In brief, the problem of the NLM was to enlist the support and participation of educated people in the programme as trainers, motivators and volunteers; enhance the performance of the existing delivery system through improvement in the quality of services to attract learners; and better-trained literacy instructors/supervisors and efficient management (resource utilisation). It was also necessary for NLM to become more sensitive to the problems of the field and for the internal communication system to become more responsive to the needs and requirements of the field.

It is against this background that NLM had to develop a communication strategy that focused on women as new learners and change-agents, engaged and mobilised the community to participate as volunteers and teachers, and created a positive environment wherein literacy (even for adults) was perceived as a necessary skill in the modernisation process.[6]

It is necessary to mention at the very outset that communication support does not refer only to the products or media materials (posters, booklets, films, slides, films and videos) but also includes consideration of the process by which such products are developed. For instance, it is possible that a literacy instructor may develop a set of flash cards to be used in class with the participation of the learners. The preparation of the materials through the participatory process is a learning experience that improves participation in and awareness about the functional literacy programme among the learners. Generally, this process is regarded as a 'demystification' process by which the learner and instructor (applicable at all levels) share their common pool of knowledge and experience gathered individually over time and collectively evolve to a higher plane of understanding/awareness and skill. This is the essence of a participatory learning process for which interpersonal communication (with or without media support) is the key. Unless there is an acknowledgement of equality between the participants at a human level, obstructions/attitudinal barriers will block the adult learning process, particularly in the case of poor women and SCs and STs who suffer from a sense of historical injustice that Paulo Freire has called 'the culture of silence' (Freire 1972a).

In keeping with this perspective, where winning the confidence of the adult learner and establishing a relationship of trust between the learner and instructor was the key objective, the role of communication was broadly grouped under the following heads:

Advocacy/Public Awareness: It was necessary to create a favourable climate for adult literacy such that the objectives of NLM were accepted and supported by all sections of society. Cynicism and doubt had to be dispelled with cogent reasoning based on sound economic, political, cultural and human arguments. Mass media, small-group media using audiovisual means, print publicity and direct mail, apart from discussion and dialogue, were required for this purpose. Also, the public awareness campaign had to cut across all economic classes, social groups, business organisations, different government departments, urban and rural milieux, students and teachers, women and men.

Motivating the Learners/Instructors: Adult literacy was a difficult task and required tenacious effort on the part of the learner and patience and skill on the part of the instructor. The need and importance of literacy had to be felt by the learner and constant support and encouragement were necessary for both partners (whether in a one-to-one situation or a classroom/centre situation). Supplementary communication materials that were visually attractive and stimulating, audiovisual and video programmes that were informative and enjoyable, games and puzzles, cultural and recreational activities that were relaxing and, above all, a physical space that was secure and comfortable, especially for poor rural women and persons from SC/ST communities, all contributed to enriching the learning situation and helped the learner and instructor in their collective endeavour.

Training in Communication Skills: A variety of communication materials were required to improve the communication skills of the instructor, supervisor and other management personnel at the district and state levels. The renewed objectives of NLM, with its emphasis on participative learning, literacy through dialect, increased use of audiovisual inputs, post-literacy materials and mission management system, had to be understood and internalised by the functionaries. In addition, the basic teaching–learning materials for the adult learner like the literacy primer required improvement and support from additional materials like charts, flash cards, slides, audio cassettes, radio, TV and video programmes.

Documentation/Learning from Sharing Experiences: Adults had a wealth of experience in overcoming difficulties, stimulating learning and innovating in field situations. This constituted a rich repository of learning 'material' in a participative and creative learning process. Such experiences had to be documented and shared widely to encourage others to think without fetters and learn from one another's experience. The documentation (written and audiovisual) also provided informal feedback to NLM so that the field process could be supported further. Success stories, problem solving skills and interdepartmental cooperation at all levels, but especially at the village, block and district levels, needed to be highlighted and shared.

The approach that NLM set out to adopt to achieve the communication objectives outlined above were:

- Energising the existing structures at the Centre, state, district, block, project and village (adult education centre) levels. The focus was on retaining clarity about NLM objectives, accountability and improved performance, faster flow of information and feedback, and a commitment to improving the quality of the programme rather than achieving notional targets of learner

enrolment. Supportive supervision with a team-based approach had to be inculcated among functionaries.
- Decentralised planning, production and effective utilisation of communication materials. However, identification of priorities, specific focus on different target groups, choice of media and communication planning had to be coordinated at the state level and between states. The Directorate of Adult Education (DAE) under NLM had to be the nodal agency to give effect to this coordinated decentralisation effort.
- Inter-departmental cooperation between agriculture, health and family welfare, rural development, women and child development and education functionaries was necessary so that field activities for social mobilisation could be planned together at the state and district levels, thereby enhancing the level of cooperation at the block and village level. This was indeed a difficult task but had to be done.
- Intensive effort in particular areas had to be initiated with the objective of complete eradication of illiteracy in selected districts, especially in states that showed a weak performance. Use of innovative materials, flexible operations, experimental project modes and careful planning and monitoring were the salient features of such intensive effort.
- Integrating a quick method of evaluation and mid-course correction in any effort was another crucial aspect of the communication strategy. This helped in aborting projects and avoiding wastage as well as in intensifying successful efforts and planning for the spread of such efforts. The key to achieving this was the speed with which such evaluations could be executed through the existing internal communication system such that the response could also flow through the same structure quickly.

WOMEN AS LEARNERS

The New Policy on Education (NPE) initiated by Prime Minister Rajiv Gandhi clearly articulated the need to address women's education as a means for achieving women's equality in society. This was an important statement and commitment on the part of the Indian government in the context of growing concern regarding the inferior status of women in our society. It was a disturbing truth that in spite of equality before the law, widespread discrimination against women prevailed even among the relatively better-off sections of society. Women's contribution in economic activities like agriculture was not even enumerated in data collection. As mentioned earlier (Section I), the women's movement had succeeded in drawing attention to the prevalence of gender-based discrimination in terms of economic and social opportunity. The Report of the Working Group on the Status of Women set up by the Government of India had noted the various discriminatory practices and recommended several measures to correct this imbalance. Women's education was perceived as one important step in this direction and the NPE, with its focus on Education for Women's Equality (EWE), reconfirmed the government's commitment (Government of India 1986).

However, there was a strong body of opinion that believed that the prevailing system of education was not suitable for the rural poor and women in particular. They argued that the government's literacy and general education programmes had led to the 'domestication of the people' rather than their empowerment and consequent liberation from oppression and injustice. This was because the content of education did not question the unjust and exploitative structure of society but reinforced the status quo and, with it, the dominant ideology. Most literacy primers did not reflect the lives of the poor—their daily hardships, deprivation and exploitation. Instead they showed a harmonious village community where everyone lived in peace and amity. Government programmes were shown as functioning smoothly and faultlessly. Every functionary was shown to be serving the people with honesty and a spirit of selfless service that hardly was the case in reality.[7]

Krishna Kumar, scholar and educationist, wrote:

> A typical literacy primer tells the learners how one poor peasant gradually becomes prosperous by making certain rational decisions, such as the decision to plan his family and to start a new method of cultivation. The gist of such narratives is that a man changed his economic condition by dropping a set of backward and disabling characteristics, and by adopting an alternative set of characteristics that were modern and healthy (Kumar 1982).

This kind of presentation was consistent with the dominant paradigm (refer to Section I) that held the view that mere information and a vision of modernisation would be sufficient to bring about change.

As we have noted earlier, this did not happen in traditional societies and literacy, which had a direct bearing on the modernisation process, remained abysmally low. Krishna Kumar (1982) went on to argue:

> A realistic literacy curriculum for adult literacy cannot be developed without acknowledging the presence of deep-rooted injustice and conflicts in rural society. How can a programme that hides so significant an aspect of the life of poor peasants and landless capture their imagination? Literacy classes have a poor response because the curriculum taught in these classes does not relate to the life of the learners.

An analysis of the primers being used for the women showed that they perpetuated existing stereotypes, restricting women's concerns and activities to housework, child care and family planning. The focus of the primers was on viewing women mainly as mothers, as being important only because they gave birth to and looked after children. Their role as productive workers in rural (and urban) society was completely ignored. Widespread male domination, the double burden of work in the fields and at home, the different kinds of social and economic discrimination against women were not even mentioned in the primers. As one commentator put it:

> There is talk of places of pilgrimage. All these are considered more important and necessary than information on the legal rights of women and social legislation on the minimum age of marriage, divorce, dowry and inheritance, equal wages etc. In the name of education and development, the primers actually promote what are considered to be anti-women notions and attitudes (Bhasin 1984).

The primers did not initiate discussion on issues of concern to women. Information and messages were delivered as 'pre-packaged truths'. The learners were not encouraged to ask any questions, to challenge existing notions about the role and position of women in society, or to form their own opinions. Literacy programmes using such primers only helped to perpetuate women's silence and subordination.

The methodology of the prevailing adult literacy programmes was also questioned. The classes in the prevailing programme were conducted like classes for children with a one-way flow of information, knowledge and ideas. The adult learners (poor rural women) were regarded as ignorant and likened to empty pitchers that needed to be filled with ideas and information. Paulo Freire had referred to this kind of an approach as 'the banking concept of education' (Freire 1972b). An alternative methodology that encouraged 'critical reflection' and fostered a questioning attitude was thus required for adults, especially for those who lived in deprivation and oppression. Literacy had to have an agenda that enhanced their awareness and gave them requisite skills to obtain their legitimate rights and entitlements in society.

This reformed vision of adult literacy was the stated objective of NLM when it was launched in 1988. In order to attract women as learners, it was necessary to first come up with alternative materials and launch a motivation campaign that recognised women's productive role and acknowledged their participation in society as dynamic agents of change. Fortunately for the NLM, some creative thinking and innovative effort had been made in this direction prior to its launch. With the encouragement and financial assistance from UNICEF, the DAE produced a set of motivational films titled *Jaag Sakhi*, and a literacy primer *Khilti Kaliyan* meant specifically for rural women.[8] There was also a set of instructional films accompanying the lessons of *Khilti Kaliyan*. These provided NLM with an opportunity to train the adult education instructors in a new methodology and inculcate in them an attitude of greater respect towards women learners and an appreciation of women's role in society.

Jaag Sakhi comprised a series of 12 television programmes for rural women. The films followed a magazine format with 20-minute episodes, each of which contained a motivational module using a dramatised sequence; an information module giving relevant, workable information by a lawyer, doctor or some other expert; and a case study. The attempt in this series of films was to inculcate in non-literate poor women the belief that, with effort and some external support, the circumstances surrounding their lives could be altered. The case studies depicted other women in similar circumstances who were being empowered through organisational effort and collective action to change their lives. By identifying with such women the viewers, it was hoped, would feel inspired to question and critically appraise their own circumstances and take the first step towards becoming literate and coming together as a collective. The programmes carried relevant and meaningful information about health, legal and social rights, and other matters that were pertinent to rural women. *Jaag Sakhi* also suggested that literacy and participation in an adult education centre (Mahila Kendra) were positive steps in this direction.

Through the use of authentic and powerful visual images of women, the underlying subtext was to highlight the inherent strength and capacity of women, thereby countering the mainstream, stereotyped depiction of women as ignorant, weak and long-suffering. The films were telecast by Doordarshan and also screened at women's gatherings through the video.

KHILTI KALIYAN

This 24-part serial aimed at women in the age group of 15–35 years was made with the intention of encouraging them to recognise the need for literacy and the changes that literacy could initiate in their lives. The serial was based on an experimental literacy primer by the same name. It was developed for women learners and dealt with themes and issues pertaining to the lives of rural women. In the course of its effort to complement the primer, the TV serial established a link with the real problems of social, economic and political deprivation and oppression faced by women. Thus, the narrative of *Khilti Kaliyan* forced the audience to consider the position of women in society and the reasons for their unequal status.

Each episode began with a short sequence enacted in the traditional folk style featuring a typical Indian street magician and his assistant. In a major departure from the tradition, these roles were deliberately performed by women who were impersonating/acting as men. The female magician shook her *damaru* to summon her audience and draw their attention to the skit being performed. The challenge presented was in the gunnysack, which the assistant failed to open claiming that it was locked. The magician then performs her trick of opening the bag, not by uttering any magical 'abracadabra' but by reciting a litany of the real problems of lack of education. Her rhyme suggested that the key to the bag of information and knowledge lay before them in the form of literacy. The sack of knowledge could be opened with the empowering key of literacy, thereby dispelling the clouds of ignorance that had for so long darkened the horizons of their lives and everyone, irrespective of age and sex, would benefit from literacy and knowledge. With a dramatic and magical burst the sack opened and out poured a jumble of letters that played in the air before forming the words of the title—*Khilti Kaliyan*. The audience watching the skit was entranced not merely with the skit itself, but with the promise of liberating knowledge that it held. Similarly, each episode of the TV serial ended with the same song, *Aurat ki kahani* (the story of a woman). The words of the song expressed the empowerment of women through literacy and their marching shoulder-to-shoulder with men to obtain their rightful place in society. The technique used to end each episode was a deliberate freezing of the expression—always positive—on a female character's face.

This projection of a positive image of women was a deliberate and major departure from the stereotypical portrayal of women even in adult literacy materials. Ironically, the printed primer *Khilti Kaliyan* fell into the same trap of portraying women with their heads and faces covered and mainly concerned with housework and the well being of their children and family. The TV serial went far beyond this in its portraiture of women. At the same time, the women in *Khilti Kaliyan* were credible characters and their demeanour and performances were quite identifiable with women in a north Indian village. The usual dressing up for the TV camera was avoided and a 'naturalistic' style of presentation was used. The strong, positive imaging of women, even if it appeared to be a deviation from reality, helped viewers in reflecting on the issues raised and the resolutions found in the film. In the evaluation study done by CENDIT for UNICEF/DAE,[9] it was found that the younger women in the sample villages identified closely with the characters and saw them as their inspirational role models. The older women disapproved of their forthright manner and outspoken dialogue but agreed with the substance of the issues raised. Insofar as the objective of the films was to raise consciousness among women by reflecting a reality of their lives back to them so that they could critically appraise their circumstances and overcome them (as was shown in the films), they succeeded in generating a conscious thought process that was necessary before meaningful change could be brought about.

At the same time, the films did not become didactic or dull by portraying women whose attributes of strength and confidence were unreachable by village women. The women in the films worked in the fields, fetched water, tried to make ends meet, suffered from poor health because of malnourishment and repeated childbearing, bore their misery bravely and worried about their children and family. The characters portrayed suffered oppression but they were helped by those who had managed to break free and in the 'space' provided by the literacy centre discussed their problems and social–economic issues and arrived at their own solutions, in the home, the workplace and society, albeit slowly and with effort and struggle. The weak characters—whether the elderly *dai* (midwife, *Sukhiya*) who learnt new skills and became a trained *dai,* or Ram Piyari who struggled to remember the letters and words in the literacy prime—did not lose courage and were helped by others in the village.

Another element in the film that was liked by the audience was the use of songs, especially in the earlier episodes. Apart from the song in the title sequence that worked as a signature tune for the TV series, there were several songs that touched a chord in the audience. The song on the *bari ladki* (elder girl child/daughter) corresponding to the fifth lesson in the literacy primer (see Figure 6.1), described the unfortunate plight of a girl child who was weighed down with the responsibility for sibling care and helping the mother with housework and had no time to play, study or just enjoy the freedom of childhood. This particular song, which ended with the depiction of the young Murti who could not visit her friend because she could not be spared from housework while her brothers were free to play, never failed to bring tears to the eyes of women who watched the film, including those from better-off circumstances because discrimination against the girl child was equally common in middle-class households too. The song that encouraged Ram Piyari to try harder to learn celebrated the mental capacity of human beings and reminded her of the number of things in nature that she could remember and recognise using just her eyes was a simple evocation of our innate capacities. The Bengali song about the loneliness of the child-bride who, torn from her family at an early age, pines for her home and pleads with the boatman to take a message to her family to come and take her home or the song celebrating the victory of the struggle of villagers to draw water from the high-caste well were all very moving experiences for the women viewers.

The entertainment programme to mark the launch of the *Mahila Samiti* in the village had a puppet show with the puppets portraying the plight of a woman who was thrown out of her home by her husband and his family. Bereft and utterly alone, she drew courage from the marching procession of women who were organised as a collective and she picked up a placard and joined them looking forward to the future. These songs not only entertained the viewers but also carried the narrative forward and sent the message home to the audience helping them cognise their own reality in terms of that depicted

LITERACY AND EMPOWERMENT

Figure 6.1
Page from Khilti Kaliyan Primer (Lesson on Bari Ladki)
Source: DAE/UNICEF.

in the film. If *Khilti Kaliyan* attempted to broaden the literacy process to include creating awareness among rural women about the circumstances of the learners and helping them through reflection to work towards altering them, the songs of the TV serial gave them the courage and fortitude to engage in the struggle to change their lives.

The TV serial also supplemented the literacy primer by showing innovative ways of teaching literacy and demonstrating that adult learning could also be fun. Instead of just using a rolled-out blackboard or slate and chalk, mud and sand, charcoal and ash, grain and dough, sticks and beads and a variety of readily available materials were used in imaginative ways and woven into the learning situation effortlessly to demonstrate that learning could be an enriching experience for rural women. These examples of teaching methods were also a hint to the literacy instructors to use their latent creativity to make the learning process more interesting. Similarly, rather than promoting passive acceptance, discussions around the lessons encouraged a questioning of views based on the experience of the learners in the literacy centre. Many opinions based on what was a desirable norm were not available to the poor and the film showed the frustration and anger of learners at being told about things that were unaffordable (like a nutritious diet for a pregnant mother). Through lively discussions, it was possible to communicate and underline the importance of the points made without negating the women's felt experience. The value of discussion in the class, an essential ingredient of any adult learning situation, was thus reinforced through the serial.

Similarly, the formation of the *Mahila Samiti* to help find an economic activity that would help them supplement their family income, give them the experience of working together, teach them to keep accounts and share and provide support to other members in times of difficulty and work together as a collective force against the social and economic discrimination against women was part of the larger adult education agenda that could not be reflected in the literacy primer. However, the *Khilti Kaliyan* TV serial effectively depicted these features in the films in a manner that was credible and feasible and related to the experience in many rural areas, especially those surrounding large urban centres. In the survey done after the telecast of the serial, many women found these aspects of adult education more interesting than literacy and wanted to initiate similar activities in their own environment; in fact, some of them had already initiated such group-based activities. *Khilti Kaliyan* made a consistent appeal for literacy as repeated in the signature tune of *Aurat ki kahani* and emphasised its importance as a necessary tool for the empowerment of women, even if it was not sufficient by itself. Becoming aware of their circumstances, critically reflecting on them, organising themselves as a collective and initiating economic activities to find their partial economic freedom were all messages woven into the narrative. Whether it was fighting against caste taboos like drawing water from a well traditionally reserved for the upper castes or fighting against the malpractices of the labour contractor who denied the illiterate poor their minimum wages, the films showed that literacy gave women the confidence and knowledge to challenge the oppressors and fight injustice to obtain their legitimate dues. In that sense, the films were a powerful reminder to all viewers that literacy was a powerful instrument for bringing about social change.

The adult education centre shown in the film was also instructive. At the beginning of the film, the instructor has only one enthusiastic young learner. But slowly the numbers grow and to retain their interest, the instructor has to deploy a variety of interesting teaching methods. The instructor also has to participate in the lives of the learners and advise them on how to cope with their problems and yet make time for class. Some of the literate women in the village support the adult education centre by persuading others to join the class or by making educational aids and toys that help learning. A major push comes with the formation of the *Mahila Samiti* to make garments to be sold in the city. The possibility of engaging in an economic activity enhances women's motivation to join the class and their numbers grow. Learners, especially the older ones and the younger, more enthusiastic ones, become creative and produce charts and drawings to enrich the learning environment and enhance the learning experience. Gradually, the centre becomes a vibrant space for the women to enthusiastically participate in creating a new life for themselves through literacy, empowerment and collective action. This is the transformative process implicit in any good adult education programme and the TV serial *Khilti Kaliyan* succeeded in portraying it in a creative and credible manner.

Though the characters in *Khilti Kaliyan* were varied, women were generally portrayed as positive and desirous of change in their status and circumstance. Many of the characters evolved through the narrative development of the serial while others gradually faded out after playing their part in the initial episodes. The main protagonist, Murti, is a typical *bari ladki* of a household who has to help her ailing mother with household chores and does not have time for school or play like her brothers. The parents are persuaded to send her for adult education classes and without neglecting her housework, she begins to learn quickly. Once at the literacy centre, her talents flower. By the end of the film she has gained in confidence and is ready to start her own adult literacy centre. Her learning comprises not merely the letters of the alphabet but also understanding the circumstances of the labouring poor and the need for them to show solidarity and stand up against oppression and injustice and fight for their rights and entitlements.

Some of the other characters like Susheela the adult literacy instructor, Usha the health worker and Kalindi the social worker from the city were strong, positive characters who had struggled for their autonomy and were helping the village women to empower themselves through education (literacy), economic activity and collective organisation to obtain their legitimate dues. Their attitude was one of warmth, respect and friendliness without any hint of patronage towards the poor or illiterate village women. The underlying 'sisterhood' theme that unites all women against patriarchy has been a key element of the women's movement and *Khilti Kaliyan* blended that very well through the characters in the serial. The other educated positive women in the village like Kamala and Ashrafi were strong positive characters who could stand up to men as equals and, with their enthusiasm, confidence and good cheer, help the others take the first steps to change their circumstances through literacy and education. The elderly *dai* Sukhiya or Ram Piyari, the elderly woman attending the literacy class, were all participants in the engaging process of change being brought about in the lives of all these women. A significant and dramatic role was that of the tailor's wife, Rasheeda Bai, who revolts against her husband and comes out of her purdah to assert that she is tired of the unfair pressure of the *sarpanch* (village elders) and *thakur* (landlord) and promises to teach the women of the *Mahila Samiti* when the tailor Umar Ilahi refuses (or goes back on his promise) to teach the women stitching and tailoring for fear of reprisal from the *sarpanch*, *thakur* and the city-based middleman.

The audience could easily identify with the characters, who were credibly portrayed, complete with contradictions and pressures. Even the *thakur* and his crony, Kamta Prasad, the *sarpanch* of the village, were not shown in the melodramatic manner usual in films. Oppression was presented as a fact of life and the women in *Khilti Kaliyan* resisted and overcame it through concerted action, thereby giving the message that change is possible and within the grasp of every individual. Other men, who may have been positive and strong in their work situation but were inconsiderate and oppressive in the home, were gently nudged towards greater sympathy and understanding for their wives and daughters. New ideas like health care and nutrition were introduced in context and discussed with respect for the traditional views. Similarly, old ideas of disease and cure or social customs like dowry were discussed in the adult literacy centre in an open manner, presenting different points of view without a prescriptive imposition from the outsiders.

The language spoken in the film was easily understood by women, particularly in north India. The film used simple Hindustani, which is the spoken language of ordinary folk and rich in oral metaphor, communicating its ideas un-self-consciously. It avoided the usual patronising manner of well-meaning government information films made by the urban middle class without any proper understanding or experience of the social milieu of the expected viewers. *Khilti Kaliyan*, on the other hand, used language with a felicity which was remarkable in that viewers could identify with the characters and storyline completely.

> The respondents easily identified with the situations and characters in the film. They were also comfortable with other elements in the film—the music, the humour, the language and dress. These elements were so characteristic of their own lives that many expressed pleasure and surprise at the fact (Kapoor and Unnikrishnan 1990).

It was in creating a credible reality for the rural viewers that *Khilti Kaliyan* was most successful. The film reflected the drama of their daily lives, their joys and sorrows, their struggles and despair. The characters, the social tensions, the fears and anxieties, hopes and aspirations came through in a manner that was identifiable by the viewers.

> In every group of women there was an eldest daughter; an overburdened housewife; a girl whose childhood had come to an end; an old woman who feared the learning process; or a woman determined to change. Each of these women saw themselves in the film and consequently there was a blurring in the distinction between the screen-image and the real-life one. On occasions the film caused a catharsis. A woman watching a scene would shed bitter tears. Another would cry out, 'But that's my story!' and, for that moment when the pain in their lives surfaced, we became the silent and helpless spectators (Kapoor and Unnikrishnan 1990).

Khilti Kaliyan was a path-breaking attempt to design a literacy primer for women and producing a TV serial that would supplement the primer was an innovative initiative taken by UNICEF and the DAE. Although made with the two main objectives of attracting women learners to adult education centres and enriching the learning process, *Khilti Kaliyan* went far beyond that in its potential 'as a radical new effort to draw women into the mainstream by transforming education into a real tool of development and change'. The TV serial was telecast by Delhi Doordarshan Kendra (DDK) once a week over 24 weeks. However, it was telecast without adequate preparation to ensure that adequate TV viewing facilities, or even the literacy primer, were available at the adult education centres. Nor were the adult education instructors trained in using the films in conjunction with the primer being taught in class. As the study conducted to assess the impact of *Khilti Kaliyan* concluded:

> From the word 'go' we found that no matter how hard we pushed we were not going to see the film being used as a literacy aid. The serial was looked upon as a welcome relief by some and a nuisance by others. Instructors merely shrugged their shoulders and kept to their own pace regardless of the literacy sections being screened. Under such circumstances, how could learners become absorbed with the serial and respond with the spontaneous keenness of women who discover that they are looking at a new version of their own life stories, a version that does not demean or negate their role in society? How could we expect them to chew over the new ideas that were presented to them when everything around them was telling them that none of this really mattered? (Kapoor and Unnikrishnan 1990).

The anguish expressed by the team of researchers is a familiar refrain in most projects that attempt to use audiovisual materials and mass media in an educational or development context. Even well-researched and creatively designed materials (like *Khilti Kaliyan*) are rendered ineffective by the lack of training and motivation of the instructors and other intermediaries, pointing to the absence of a systemic approach to the use of communication technology and materials and proper understanding and appreciation of the power and cost-effectiveness of the new media. *Khilti Kaliyan*, therefore, remains a bold experiment and a milestone in applying creativity (see Box 6.1) to motivate women towards education by giving them a broader transformative vision of the educational process. In the overall attempt of NLM to address its main clientele of women learners, *Khilti Kaliyan* focused on the reality of their lives, which has to be addressed if literacy and education is to have meaning for them.

Box 6.1
Padhna Likhna Seekho—Literacy Song by Safdar Hashmi[10]

Padhna likhna seekho, o mehnat karne walon
Padhna likhna seekho, o bhookh se marne walon
(Learn to read and write, O all those who are labouring
Learn to read and write, O all those who are dying of hunger)

These were the opening lines of a song written by Safdar Hashmi, writer and theatre activist, for the National Literacy Mission. The song was inspired and based on Bertolt Brecht's composition in the play *The Mother,* in which the labouring poor and hungry are exhorted to become literate in order to understand the causes of their misery and bring about a radical social transformation of society. *Padhna likhna seekho* became the rallying song for the NLM. The energetic and rousing beat of the music and the evocative visuals accompanying it for the television version was attractive for the young volunteers in the literacy programme.

Safdar Hashmi was a rare talent, deeply committed to fight against injustice and oppression of the labouring poor. He had formed a theatre group Jana Natya Manch in Delhi that did street-corner plays on wide-ranging themes concerned with exploitation and injustice. Some of these plays were landmarks in Delhi's theatre that was moving out into the streets at the turn of the 1980s. His plays were performed mainly in working-class colonies and neighbourhoods and as a political activist Safdar was hugely effective. He had already completed the script for *Khilti Kaliyan*, the TV serial on women's literacy that set the tone for a robust, dignified and positive portrayal of the labouring poor, particularly women and their struggle for justice and a better quality of life.

The original composition of *Padhna likhna seekho* was five minutes long, but for repeated telecast a shorter, one-minute version had also been produced. The Ministry of Information and Broadcasting and Doordarshan officials liked the TV spot and agreed to NLM's request

(Continued)

> **Box 6.1**
> *(Continued)*
>
> for free and repeated telecast of the song. It caught the attention of the audience and became quite popular. However, some viewers wrote to Doordarshan objecting to the line *bhookh se marne walon*, arguing that the line was contrary to the government's claim of having abolished starvation deaths. Because of these literal interpretations, a line that held such strong appeal for the struggling poor was lost on the more complacent middle-class audience. Ironically, many of those who objected were officials from the adult education programme who felt that the literacy programme could not make an appeal to those who were struggling for survival. The whole point made by Brecht (and Freire's conscientisation programme) of using literacy for raising consciousness about rights and fighting oppression and injustice was lost on this audience. Doordarshan officials got worried and requested NLM to withdraw the song or revise it.
>
> In the meanwhile tragedy had struck Jana Natya Manch. On 1 January 1989, while performing a play in a working-class colony of east Delhi, the team was attacked by the hired goons of local politicians and Safdar Hashmi was brutally beaten up. He succumbed to his injuries in hospital. There was outrage among the theatre workers who held a series of protests and demonstrations against the ruling government. The large turnout of workers and ordinary people at the funeral rally for Safdar Hashmi showed the love people felt for him. A Safdar Hashmi Memorial Trust (SAHMAT) was spontaneously formed to uphold freedom of artistic expression. The untimely death of Safdar Hashmi was an irreparable loss to Indian theatre and political activism through the arts. His association with NLM during the last couple of years of his life was enriching for the government programme and demonstrated that creativity and deeply-felt social concern for the disadvantaged and oppressed could lift the profile of a government-sponsored social development programme, usually regarded with scorn and distrust, to a vigorous and vibrant rallying call for social transformation.
>
> The *Padhna Likhna Seekho* song could not be revised, as Safdar was no more, so the option was to delete the *bhookh se marne walon* line and replace it with *padhna likhna seekho* for Doordarshan. It was a poor compromise, but the song remained popular and had high recall among audiences. At the field level, during the Total Literacy Campaigns (TLC), volunteers and Bharat Gyan Vigyan Samiti (BGVS)

> **Box 6.1**
> **(Continued)**
>
> activists used the song without modification as a rallying call. It was translated into several languages and used widely for many years. Some years later, during the International Literacy Day (ILD) celebrations in 1993, the original version of Safdar Hashmi with the line *bhookh se marne walon* was reintroduced at the function in Delhi before an international gathering. By that time, the earlier objections had been forgotten and the popularity of the song in the field ensured its legitimate recognition.

PROJECT IN RADIO EDUCATION FOR ADULT LITERACY (PREAL)[11]

The National Literacy Mission (NLM) was conceived as a societal mission to eradicate adult illiteracy. It was one of the five Technology Missions that emerged as priority areas for the application of modern technology and management practices to meet social objectives. Using communication technology like film, radio and television was a natural corollary for application in NLM's programme formulation.

A fundamental issue in NLM was the concern to improve the pace and content of the literacy teaching–learning process. Right from its inception in May 1988, NLM had been engaged in the endeavour to reduce the time taken by individual adult learners, particularly women, to achieve the prescribed levels of functional literacy. A new pedagogy called the Improved Pace and Content of Learning (IPCL) was developed to enhance the quality of learning materials while shortening the time-span for achieving the NLM norms of functional literacy. It was here that the application of modern electronic media became relevant. An experiment using radio as an input to intensify the teaching–learning process was conceived as an innovative collaborative project between All India Radio (AIR) and NLM.

PREAL was operational in 16 selected districts of Bihar, Uttar Pradesh, Madhya Pradesh and Rajasthan. Weekly programmes under the title *Nai Pahal* were broadcast from eight AIR stations that covered these districts. Some additional relay centres were added later to cover the dark areas where the signal strength was insufficient. The objective of PREAL was to study the effectiveness of using radio lessons to enrich the learning experience of women learners in adult education centres (AECs) and thereby sustaining their interest in attending the classes regularly and achieving NLM's prescribed literacy norms. Particular emphasis was laid on reinforcement of reading ability through a planned and systematically graded reading drill that was inducted into every lesson that was broadcast. The instructional content was in standard Hindi but the spoken dialect of the particular regions was also used to enrich programme content, vocabulary and cultural specificity. In tribal districts, however, the instructional content followed the NLM norm of bilingual primers, that is, to initiate literacy in the local tribal language and vocabulary and then gradually build up to standard Hindi. Five hundred AECs in non-tribal districts and 300 AECs in tribal districts were identified for each AIR station, making a total of 3,800 AECs. Presuming that 25–30 women would assemble at each AEC, the project was reaching out to 95,000–114,000 women over a period of six months through 26 weekly lessons broadcast from the participating stations with one repeat broadcast every week. The experimental nature of PREAL was further strengthened by the research objective of the project where the National Council of Educational Research and Training (NCERT) and ISRO's Development Education and Communication Unit (DECU) were the collaborating agencies for the design of the software's instructional content and the formative research and summative evaluation of the project.

Identifying AECs for the project was the responsibility of the SDAEs in the four Hindi-speaking states and training to supervisors and all adult education instructors was imparted through State Resource Centres (SRCs). An additional hardware input of two-in-one radio-cassette recorders was supplied to all the 3,800 AECs in the project along with batteries and cassettes for recording and later playback to those learners who might have missed the broadcast. A specially designed supplementary radio reader titled *Akashvani Pathmala* (Figure 6.2) containing the literacy (reading) material was supplied to all learners. All the selected instructors were trained to use the hardware and software (radio programmes and radio reader) through a specially developed training manual and some radio programmes developed for this purpose.

Figure 6.2
Akashvani Pathmala

The most important part of PREAL was defining the objectives of the software design and articulating specifically the role expected from radio and how that would be reflected in the programmes. The priority geographical area of the four Hindi-speaking states and the target group of women learners were already decided for PREAL. Within this framework, in January 1990 an attempt was made during a PREAL planning workshop in Jaipur to define the scope of radio in teaching literacy. The main point that emerged was that literacy was more than a mere mechanistic skill. It implied the facility to read and understand the meaning of the written word that would lead on to further action. In other words, literacy would require the development in learners of the ability to think critically and independently and take decisions regarding action. The stages by which such development would take place were identified as 'developing the ability to listen with attention, speak with clarity, read with understanding, think with independence and write with confidence and conviction.' If literacy was regarded as a skill that encompassed this enlarged spectrum, then the radio programmes had to be integrated in the teaching–learning process in the AECs and the instructors trained to use them. The radio lessons had to become a catalyst to transform the process of literacy instruction in the AECs. The essential objective of the radio programmes was to intensify the learning process through rich cultural and entertainment content rooted in the local milieu and an instructional design that helped develop the ability to listen with attention, recall and analyse the content of the programmes, engage in the reading drills and verbal exercises as well as internalise the basic messages on the relevance of literacy in the life circumstances of women learners.

In order to avoid variation in the reading drill (the core literacy content), the following pattern was adopted by all radio stations:

Suno aur Bolo (listen and speak): First, the radio presenter reads while the AEC listeners listen with the radio reader *(Akashvani Pathmala)* open on the correct page. Then the radio presenter and radio learners speak together and the AEC learners also repeat along with them.

Suno aur Dekho (listen and see/follow the text): Only the radio presenter reads the same lesson while the radio learners and AEC learners follow the written text (words) visually in the reader *(Akashvani Pathmala)*.

Dekho aur Padho (see/follow the text and read aloud): In this sequence, the radio presenter introduces the activity to be undertaken, that is, looking and reading (deciphering and reading aloud). On radio, the radio presenter and radio learners would read the whole exercise.

A clear cue (easily identifiable sound) was given every time the learners in the AECs were expected to repeat anything spoken on the radio. The AIR producers were also instructed not to curtail the last sequence of *Dekho aur Padho* though they were given the liberty of reducing the other two sequences in case there were time constraints. As the radio lessons progressed the pace of reading and repetition was increased on the assumption that the learners in the AECs would have become more familiar with the alphabet through practice.

The last few lessons were common to all stations, including Indore and Ranchi where the tribal bilingual reader was used, as they were just stories read out by the radio presenter and, if time permitted, repeated by the AEC learners. This was based on the assumption that by the twentieth lesson, the learners would be able to not only read but also discover the pleasure of reading and enjoy it for its own sake.

AIR also planned an interactive capsule of 10 minutes that followed each *Nai Pahal* broadcast of 20 minutes. In this capsule, production teams from AIR would visit AECs and speak to the learners and instructors and these interactions would be edited and broadcast after each lesson. A separate interactive capsule was planned for the weekly repeat broadcast of the radio lesson.

As stated earlier, PREAL was a collaborative project of NLM and AIR, with DAE serving as the executing agency. The active participation of several other agencies like NCERT, DECU, UNICEF, the Central Electronic Engineering Research Institute (CEERI), SRC, the SDAEs and District Adult Education Officers (DAEOs) was crucial. The interdependence of each component on the satisfactory performance of all other components was emphasised. Care was taken to identify each contributory factor so that the broadcast of radio lessons was utilised to the maximum. The hardware, supply of batteries, additional print materials (radio reader and training manual), training and a system of monitoring were taken care of. There was also sufficient flexibility in the project to apply mid-course corrections like changing the date of broadcast, including other AIR relay stations to improve signal strength or altering some lessons from one of the AIR stations, etc. While the coordination and planning at the central level was good, its effectiveness at the field level was missing. Inflexible administrative procedure and lack of commitment combined together to render the plans ineffective at the field level and additional funds made available for PREAL remained unutilised.

PREAL also suffered because it was mistimed. It had adopted the centre-based approach to adult education with learners assembling at a one place as a group and participating in a literacy class. During the life of PREAL (1990–91) NLM's programme strategy changed to a

Total Literacy Campaign (TLC) approach based on the success of the Ernakulam (and later Kerala) model of social mobilisation and volunteer teachers. Though the TLC did not get adopted in the Hindi-speaking states till much later, it gave the wrong signal to the adult education instructors and supervisors. Word got around that the AECs would be closed down and, consequently, the already weak programme became almost non-functional. Even so, where the instructors were motivated and had internalised the value of PREAL, they took the initiative to overcome many of the obstacles and managed to conduct regular classes, including listening to the radio broadcasts to the benefit of the women learners.

PREAL was affected by the prevalent political situation in the country as well, particularly in north India. It was conceived at the high noon of Prime Minister Rajiv Gandhi's Technology Missions initiative. By the time the broadcasts commenced, two governments had changed at the Centre and power had changed hands in the states. The anti-Mandal Commission riots and later the *Ram Janmabhoomi Rath Yatra* created instability and insecurity with transport and communication being disrupted and work in offices slowing down. The project operated when the country was in the throes of a deep crisis—political, social and economic.

Yet, PREAL was a learning experience and its documentation pointed to the careful planning and preparation necessary for such a project. The producers took into account the formative research that had been undertaken before the project was launched and care was taken to relate the radio lessons to the daily life circumstances of women learners. The feedback study[12] concluded that the programmes 'were interesting, held the attention of the listeners and were found to be useful in the teaching–learning process'. The broadcasts were regular, the reception was clear and the arrangements for listening were satisfactory. The radio programmes followed the pedagogy of *Suno–Dekho–Padho* systematically and the learners, by and large, followed the instructions of the radio programmes. Both learners and instructors found the programmes easy to understand, liked the format and presentation style and found them useful in teaching–learning (see Figure 6.3).

While all this was positive, the AEC itself did not function regularly. Sometimes the instructor was not present and at other times the learners were not there or the two-in-one sets had problems or the batteries were weak. The organisation and management of listening

LITERACY AND EMPOWERMENT

Figure 6.3
The First Lesson of PREAL

sessions at the AEC were also poor and therefore exposure to PREAL broadcasts was not regular. Consequently, the effectiveness of PREAL in terms of reinforcing reading ability was limited. In conclusion, it can be said that the management of PREAL was weak in comparison

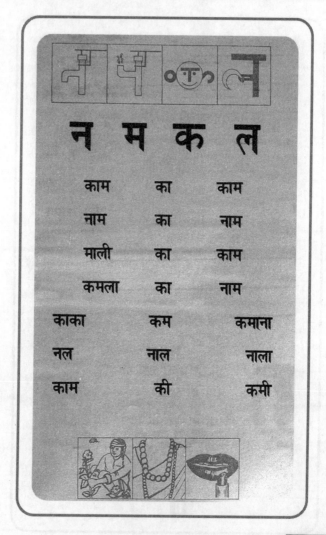

Figure 6.4
Chauraha TV Lessons

to the magnitude and complexity of the project. The decision-makers in the government, both at the Centre and at the state levels, did not fully appreciate the scale or significance of the project. Extraneous factors like the overall insecure socio-political conditions prevalent at the time also affected the project adversely.

The question therefore was how to avoid similar pitfalls in the future. The use of electronic media, which involves sophisticated technology and high cost of production, distribution, transmission and reception, requires a sufficiently large-scale operation to even test the experimental design. In order to obtain that level of infrastructure and facilities (in the social sector), the involvement of the government machinery is inevitable. Immediately, the problems of procedures and delays, inflexibility and indifference are encountered. To win the battle of motivation, to make government functionaries believe that things can be done differently is almost impossible. It is noteworthy that the SITE project was operational during the Emergency of 1975–76 and the central authority could ensure that orders were executed down the line. That scenario was no longer available and projects would have to be conceived on a more decentralised and manageable scale. The problem of such a model is that the expertise required for pedagogy, research design, training and software is not easily available in all states and districts. Had the states taken the initiative, the advantages of the well-researched software and training materials, as well as the hardware, could have been used later for a better-managed repeat broadcast of the PREAL lessons. Unfortunately, NLM's vision had shifted completely to the TLC model and no such effort was made to use the software again. Similar was the fate of the instructional TV series, *Chauraha* (see Figure 6.4 and Box 6.2).

Box 6.2
Chauraha—An Instructional TV Serial

Chauraha was an ambitious project of NLM. This TV serial attempted to teach reading and writing the Devanagari (Hindi) script. It was based on the belief that instruction through a powerful audiovisual medium like television would quicken the pace of learning and adults could be made literate in a shorter span of time. *Chauraha* was a set of 40 15-minute TV film episodes that, for the first time in India, used sophisticated computer animation techniques to teach Hindi writing within the overall framework of a narrative storyline. The technique was to show an easily identifiable image from daily life (or a graphic

(Continued)

Box 6.2
(Continued)

representation) and then superimpose a letter that could be associated with it. For instance, the first lesson used the image of a village hand-pump (or *nal* in Hindi) and linked it to the Hindi letter 'na' that resembles the shape of the hand-pump. The idea was that the visual association would help the learners to recall the shape of the letters, thereby making it easy for them to write the letters.

The storyline of *Chauraha* followed the pattern of a TV serial filled with emotional content as the main characters went through their travails in life. Its theme was woven around the value of education. Apart from the use of computer animation for the instructional part, *Chauraha* also introduced three large muppets—Panditji, a dhaba owner, Rahim Khan, the postman, and Tarseem Singh, a truck driver—who interacted with the live characters and delivered messages of harmony and cooperation, conservation of the environment, and information on the cultural variety and heritage of India. An animated pencil was introduced as another character that helped the learners to write the letters of the Hindi alphabet. *Chauraha* combined direct instruction with awareness on various development issues and did so in an entertaining and enjoyable manner. Sometimes this kind of a television format is referred to as the 'enter–educate' format (see Singhal and Rogers 1999) that uses the narrative to attract audiences to serious issues through their involvement in the evolution of the characters in the film.

Chauraha was designed as an intensive daily telecast to the Hindi-speaking areas through Doordarshan. Spread over two months (eight weeks), *Chauraha* was telecast five days of the week (Monday to Friday). Every fifth episode (telecast on Fridays) was a repeat lesson in terms of the letters introduced. It was understood in NLM that *Chauraha* would not be a stand-alone instructional TV serial and would require some supplementary print materials and the TV viewing situation would have to be facilitated by an adult education instructor. Accordingly, a *Chauraha* reader was prepared using the same sequence and visuals used in the TV serial. The adult education instructors were provided with large charts for every *Chauraha* lesson to help the learners assembled at the AEC to practise writing.

As a pilot experiment,[13] before the commencement of the telecast in 1991, 50 adult education instructors in Delhi were trained to use the *Chauraha* serial and print materials in the classroom in a

Box 6.2
(Continued)

non-broadcast mode. NGOs were responsible to ensure that TV sets and VCRs were available in the AECs and the programmes were screened daily following the same pattern as expected in the telecast. The objective was to assess the effectiveness of the learning package and the gain in the pace of learning through the application of the new audiovisual medium of television. While the study did show positive results—the instructors found the materials useful and the learners liked the format of the programmes and found the animation attractive—the usual problem of sustaining learner motivation to attend classes daily remained and the women would often miss a lesson which the instructors would then have to repeat.

While it was expected that NLM would have the time to prepare the ground by informing the Hindi-speaking states about Doordarshan's telecast schedule and ensure that there would be some provision for regular TV viewing, Doordarshan began the telecast without prior intimation to NLM. Since it was an intensive telecast schedule, the AECs in the states, even on a selective basis, were not adequately prepared to ensure regular viewing. The entire telecast was finished in two months and its impact was minimal. Ironically, because it was a national telecast, children and educated persons in other parts of India who were unfamiliar with the Devanagari script found the telecast useful for learning the script!

The lesson from the *Chauraha* experience was that planning and developing good quality materials were not sufficient for cost-effective application of communication technology using a sophisticated medium like television. Preparing the ground, ensuring availability of the hardware, sustaining learner motivation, providing supplementary print materials, training the instructors to use the materials and design other learning activities had to be an integral part of the planning process. Coordinating with the agencies responsible for the telecast (Doordarshan) and the state-level agencies responsible for training and field-level management were crucial elements missing in the planning of *Chauraha*. The possibility of developing an innovative set of materials using sophisticated technology was too attractive and it was mistakenly thought that the other components would fall into place. The magic wand of television, it was hoped, would overcome all obstacles. The reality turned out to be quite different.

(Continued)

> **Box 6.2**
> **(Continued)**
>
> It is also sad that the *Chauraha* films and print materials were not used later in a non-broadcast mode. At least that would have extended the use of *Chauraha* beyond the single telecast by Doordarshan. The only consolation was that the computer animation was so attractive that the Central Institute of Education Technology (CIET) responsible for educational TV for schoolchildren later used these excerpts in its daily telecast as a playful reminder to children on correct Devanagari (Hindi) letter formation.

A PEOPLE'S MOVEMENT FOR NLM: LOOKING FOR VOLUNTEERS

When NLM was launched in May 1988, there was very little conviction about the ability of an adult literacy programme to give 'a second chance' to adults to become literate. The existing adult education programme, launched 10 years earlier in 1978 with much fanfare by the Janata Party, which was then the party in power, had gradually dwindled to a very weak programme. Most decision-makers felt that the focus should be on universalising primary education. While it is true that primary education has a much higher claim on society's investible resources, an adult literacy programme based on voluntary effort and high degree of motivation can provide complementary support to the primary education programme as non-literate parents value education for their children more as they become literate themselves.

It was to the credit of the Rajiv Gandhi government, with its New Policy on Education (NPE) and focus on women's education and adult literacy, that NLM was included as a Technology Mission with a time-bound framework and specific goal. NLM was given a target of making 80 million persons in the age group 15–35 years literate, according to specific norms, by 1995 (later changed to 100 million by 1997). The quality of materials was enriched through the new graded curriculum, known as the Improved Pace and Content of Learning (IPCL), designed to shorten the learning cycle to 150 hours.

The method of instruction was made more robust with the use of audiovisual materials and other creative and participatory adult learning tools. However, given the general cynicism and apathy prevalent in society, there was a need to 'generate demand' for adult literacy. Hence, there was a search for a communication strategy that would position adult literacy on a higher level of priority and motivate learners to come forward and enlist for the adult literacy classes. A large-scale programme of adult literacy also had to depend on volunteers to teach the non-literate persons.

The Technology Missions, including the NLM, took a professional, management-oriented approach to achieving their objective and the support of a professional communication agency outside the government was sought for creating an environment that would enable the NLM to find educated volunteers and motivated adult learners. With the assistance of UNICEF, DAE engaged Ogilvy & Mather, a reputed advertising agency to design a communication strategy for NLM. The brief presented to the agency included the instruction that NLM's objective was to be achieved by generating a 'mass voluntary movement' in the country, that is, by creating a 'people's movement' that did not rely on the government as the sole catalyst. The route suggested by DAE to the agency was to build a positive value association with literacy; to prepare and motivate learners to accept the learning process; and to motivate the educated to offer themselves voluntarily for the each-one-teach-one or the teach-few programme of adult literacy instruction.

Recognising that the communication exercise could only instigate a process and the fact that the reach of mass media was restricted to mainly urban areas, three metropolitan cities and urban areas in four states—Uttar Pradesh, Rajasthan, Orissa and Gujarat—were chosen for the exposure. The research team travelled into the districts of Sabarkantha in Gujarat, Unnao and Barabanki in Uttar Pradesh, Sikar in Rajasthan, and to the cities of Ahmedabad, Lucknow, Aligarh, Jaipur, Bombay and Delhi. The team spoke to as many people as possible—personnel at the SDAEs and SRCs, faculty members of University Adult and Continuing Education Departments, voluntary agencies and youth bodies, and instructors, trainers, supervisors, adult learners and potential learners who had yet to enrol. The research team attended day and night adult education centres and training programmes for instructors and supervisors and also met village leaders and ordinary people. Through this field research, the team came to understand the method by which the adult literacy movement had to be generated. A mass 'people's movement' had to ignite itself within a short span of time—three to six months at the outside. Once the momentum built up, the perceived value of literacy had to be quickly felt and disseminated, with the delivery system responding to all enquiries and offering active encouragement and guidance during the 150-hour learning process. Otherwise, the heightened energy that was created would dissipate rapidly.[14]

In the absence of a charismatic national leader or the transformative historical context of a political and social revolution or newly attained independence from colonial rule (as happened in other countries), the ignition for a mass 'people's movement' could only be provided by an individual who responds to a call. The movement would then grow from the conviction and belief of many such individuals who influence their environment and their group, which in turn would change a community, create ripples in a state and finally galvanise a nation. This was the effort in NLM regarding the route the adult literacy movement should take in India, with its momentum spurred by voluntary effort rather than relying on the government as its catalyst.

The field visits also refined the key target groups among the illiterate and neoliterate people. The general category of urban workers, women and poor, rural farmers and agricultural labour with particular focus on SC/ST was modified with a different order of priority.[15] The first priority group comprised unmarried women (including SC/ST) in the age group 15–17 years, married women in the age group 15–20 years, and married women of 20 years and above. The reason for segmenting women by age and marital status was that the motivation for an unmarried girl to become literate was the higher value placed on a 'literate *bahu* (wife)'. On the other hand, the motivation for a married woman was to give her children a better future by improving her knowledge base through literacy. A further reason for subdivision by age was because (sadly) by the time a woman reaches 20 years she is a mother of two or three children and believes that her 'life is over'. So her motivation to become literate is very low.

The second group in order of priority was men (including SCs/STs) in the age groups of 15–20 and 21–35 years. The reason for this was that younger men still aspire to do better while men over 21 years are fatalistic and negative in their approach and therefore unlikely to accept a literacy programme. The exceptions were SC/ST men who perceived literacy as the means of 'release from their conditions of discrimination and deprivation'. In addition to these two priority groups, the research team recommended the inclusion of children between the age of 9 and 14 years who were out of school (never enrolled in school or dropouts) because otherwise the number of illiterates in the age group 15–35 years would increase faster than the numbers made literate under NLM's programme.

The research team also observed that the learner's interest lay in skill development with an eye to improving her economic status and reading, writing and numeracy that helped in that was all that was desired. However, while the adult literacy programme concentrated on imparting the three Rs, the national objective was to build awareness of and motivation towards economic, social and civic advancement. The value that literacy offered to the illiterate was to raise knowledge and awareness so as to prevent exploitation—both social and economic. Literacy also equipped a person to lead a better or improved life by not being dependent on others, to have access to higher earning opportunities, to take care of one's children, and so on. Since this aspect was more positive and aspirational, it was decided that it be given more emphasis in the communication.

Again, it was evident from the research that those who currently offered themselves to the adult education programme, either voluntarily or in a paid capacity, more often than not had a background of National Cadet Corps (NCC) or National Social Service (NSS) type of 'community service'. There was also a high degree of frustration among adult education functionaries because of the 'negative political and bureaucratic environment' as well as the absence of any perceivable/tangible result or benefit of the adult education effort. Hence, the role or the task before the communication effort was defined as 'creating an environment where literacy *assumed* top-of-mind priority for every individual' and became the foundation for solving every other ill in society. It was therefore clear that the adult literacy programme was not a movement for illiterates. In fact, it was a 'Mission for the Literate' as it was this group that would make adult literacy a mass voluntary movement in the country.

Having identified the key target group and the revised task of communication, it was necessary to identify the key motivation that would spur a literate 'volunteer' to action. The current perception of this group was that illiteracy was too large a problem—a perception compounded by ignorance about the 'adult education' programme and a reluctance to deal with the government. The working hypothesis, based on the field research, for the team was that in many people there was a latent desire to do something for the larger good of the community or society—the motivation of service or *seva*. Various creative approaches were developed on this hypothesis and these were tested to check the factors that would motivate the literate to volunteer for the 'each-one-teach-one' programme and gauge the reactions to alternative creative approaches. The research was conducted at the end of July 1990 in Lucknow (where voluntary effort was virtually non-existent) to represent the worst possible scenario. Since the earlier field visits had indicated that most volunteers in the adult education programme were college students (mainly girls), the research was conducted among *three sets of women* from different income groups covering students, middle-aged and older housewives, and *two sets of men* in middle and upper-middle income segments covering young adults and retired men.

Four groups emerged based upon their attitude and belief: the committed, the fence-sitters, the apathetic, and the cynics (see Table 6.1). Not surprisingly, the second, third and fourth groups were very large in their representation. Second, the literate used the terms 'literate' and 'educated' synonymously. This lead to the perception that people, if made literate, would attain the same level as the educated and thus compete for limited job opportunities. The threat of this eventuality evoked anxiety. When the distinction between the two terms was clarified, their response was that literacy by itself was inadequate, that mere reading and writing did not provide a person with the knowledge or capacity to utilise his or her full potential. It was only education that did this and thus drew the caste and class boundaries between 'them' and 'us'. Third, illiteracy was perceived as a problem linked with the 'lower classes' or the rural areas and hence the urban literates distanced themselves from the problem.

The research also showed that the earlier hypothesis that in many individuals there is a latent desire to do something for the larger good was not true. Managing one's daily existence was the major concern and there was little time or thought for the less fortunate. It was therefore necessary to change this attitude before NLM could hope to enlist volunteers. The creative strategy was therefore revised to appeal first to the 'committed' few: namely, young students and older men, and get them to act, and then extend it to the fence-sitters and later

Table 6.1

The Search for Volunteers

Segment	Key attitude	Beliefs/Behaviour
The Committed (students, mainly girls and older/retired women and men)	'The government can open schools, pass laws, but it is up to us to bring about change'	• Idealistic • Aware and involved with social issues • Strong belief in individual effort and commitment; also to *seva*
Fence-sitters (all segments)	'Social work is important—but when I have the time'	• Pays lip-service • Aware, but involvement more intellectual than emotional • Tends to act only when there is a leader • Low self-confidence
The Apathetic (all segments)	'It's not my job'	• Fatalistic • Involved only with personal problems, not concerned with social/national issues • Believes social change is the government's job
The Cynics (adult men)	'How will change benefit us?'	• Negative • Involved with economic and social issues only to criticise

Source: Communication Strategy for NLM (1989).

to the apathetic. Also, the message had to make an emotional appeal, allowing them in turn to rationalise the matter in their minds.

The final creative approach *Chalo Padhayen, Kuch Kar Dikhayen* (teach someone, do something worthwhile) focused on 'what is in it for me?' and 'what do I feel?', not just 'what does the illiterate get?'. The treatment focused on the sense of joy and achievement of the teacher in having converted a human being to stand up for herself and be able to live a life with self-respect and dignity. To spur the volunteer's action thereafter, every piece of communication had a response mechanism through a universal and easy P.O. Box 9999 across all states. What was crucial, however, was to ensure that the letters received were responded to promptly with appropriate material and instructions so as to convert interest into action.

The campaign was created for TV, radio, press, on all postal stationery, railway tickets, posters and wall paintings, as well as for non-conventional media in local languages (Figure 6.5). The song that underscored all the television and radio spots—*Purab se surya uga/Phaila ujiara/Jagi har disha/Jaga jag sara*—captured beautifully a young mother's reason for attending night literacy classes in a village outside Ahmedabad: 'to awake from darkness into light'. Having finalised the campaign, adult education functionaries from different states (where the campaign was to run) were invited and their comments regarding the translations were incorporated. The adult education (AE) functionaries were also briefed on the necessity to be alert and be prepared to respond to requests for information and materials that may come to P.O. Box 9999.

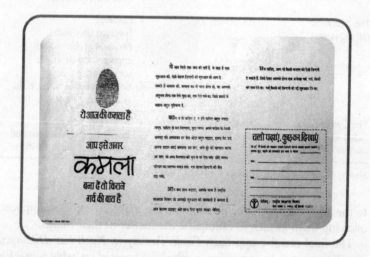

Figure 6.5
Copy of Hindi Press Advertisement for Literacy

By the middle of 1990, everything was ready for the mass media campaign to break on radio and television. However, by that time, a stipulation had been laid by the government that any public service message broadcast over radio and TV, if on a commercial time slot, would also have to be paid for. The argument was that these slots earned revenue for AIR and Doordarshan and the loss of revenue would have to be made up by the respective government departments. Since the NLM did not have the requisite resources, the entire mass media activity had to be put on hold. However, NLM was very keen to proceed with some exposure on mass media to support the literacy drives already underway in several districts of the country—such as the TLC initiated in Kerala, Pondicherry and some districts like Bijapur in Karnataka, Medinipur in West Bengal and Chittoor in Andhra Pradesh. So, with a limited media budget of about Rs 7 million allocated for the purpose, the media campaign was launched in press, radio and television in February 1991. With only limited exposure for the long creative units (70-second TV spots), the campaign hardly made any impact. There were other distractions like the dissolution of the Lok Sabha and the announcement of general elections in May 1991.

Soon after the new Congress government had taken charge after the elections, a full presentation was made to the Ministry of Information and Broadcasting on the background, task and communication strategy of NLM and the need for an intensive campaign on mass media to support the initiative on the ground. The understanding and sympathetic response of the Ministry ensured that full and regular exposure on AIR and Doordarshan could commence from August 1991.

If we consider the number of responses/letters received through the Post Box 9999 as an indicator of the effectiveness of the mass media campaign, then the *Chalo Padhayen, Kuch Kar Dikhayen* campaign was very successful. Before the intensive mass media burst, the DAE received around 30 letters per week. The day after the first telecast of the TV spots 90 letters poured in and that was when the mass media exposure was restricted to once a week in February 1991. This pattern remained as long as the mass media exposure was limited to once a week—good response immediately after the telecast and then a sharp drop. Once the spots began appearing daily from August the number of letters rose dramatically and by January 1992 the DAE in New Delhi alone had received over 150,000 letters. Most of the letters were from Delhi or the cities in Uttar Pradesh, Madhya Pradesh and Rajasthan. Fifty-seven per cent were for information, 14.5 per cent were for literacy kits, 5 per cent for guidance in literacy instruction, 7 per cent for financial assistance and 9 per cent for jobs.

Most letters were largely in response to the TV spot, while radio followed as a distant second and the long copy approach in print getting the least response. The responses at the state level in Andhra Pradesh, Karnataka, Maharashtra and West Bengal averaged 3,000 letters per month. By the end of NLM's six-month intensive mass media campaign, over 200,000 letters had been received from across the country.

Clearly, NLM's communication campaign was a success. It supported the voluntary effort initiated through *kalajathas* and the TLC in the districts and enthused the adult education functionaries and volunteers. However, the delivery system remained weak and responding to the letters took a long time or remained unattended thereby failing to take advantage of the enthusiasm generated by the campaign. Providing guidance to volunteers and encouragement during the teaching process linking literacy with the skill development needs of the illiterate were also missing. Where the groundswell was good, like in many of the southern states, the momentum generated was taken forward through the TLC. Elsewhere, in most of north India, it failed to sustain the interest of people for long.

The creative strategy for the campaign hinged on the idea of imparting to the volunteer a deep sense of emotional satisfaction (benefit) which would ensue from making another person literate and thereby changing her life. The line *Chalo Padhayen, Kuch Kar Dikhayen* (teach someone, do something worthwhile) used in all the advertising was an attempt to inspire the voluntary spirit or desire to do some service to others latent in everyone. There had been very little advertising on this issue before this and the campaign ran on radio and TV continuously from March 1991 till January 1992.

A study conducted by MARG (1992) to assess the impact of the campaign showed that 75 per cent of the audience had become aware of the campaign and to a large extent regarded illiteracy as an important social issue. About 45 per cent claimed to have been motivated to some extent to teach and over 20 per cent of all respondents claimed to have started teaching.

From the primary target audience universe of students, housewives and retired people, 19 per cent responded to the advertising and sought literacy-teaching kits. Adult literacy functionaries found greater appreciation of their work among their colleagues. The study also revealed that there was a vast pool of untapped volunteers and the delay in responding to requests for literacy kits had resulted in subsequent demotivation.

It was fortunate that the mass media campaign was running when the first success of the TLC in Kerala, Pondicherry and other districts like Pudukkottai, Medinipur, Chittoor and Nellore was making news. The combined effect was to ensure that adult literacy became a national priority with increased investment and acceptance of the TLC model as the dominant strategy. Films documenting the experience of the TLC campaigns showed the upsurge of voluntary effort in various parts of the country and kept up the motivation of the literacy workers. In terms of social mobilisation, the years 1990–92 saw, perhaps, the most successful and effective campaign ever undertaken for basic education in this country.

In contrast to this successful social mobilisation campaign of NLM in the early 1990s, a similar effort for UEE attempted later did not take off (Box 6.3).

Box 6.3

The NEEM/UEE Campaign

After the 1996 general elections, the new government at the Centre was very keen to introduce a Bill in Parliament that would make the right to primary education a fundamental right. In order to create awareness among parents, teachers, social workers and the community, it was felt that a social mobilisation campaign should be launched to support the National Elementary Education Mission (NEEM) that was to be the executive arm of the Government of India to ensure achievement of UEE.

With the support of UNICEF, a qualitative and quantitative study was conducted by Mode Research (1995) amongst parents, children, teachers, local leaders, officials and policy-makers in the states of Maharashtra, Tamil Nadu, Andhra Pradesh, Madhya Pradesh and Bihar to aid in understanding the attitudes of the various target groups. This study was the key input in designing communication interventions and also provided the benchmarks against which the results of the interventions could be evaluated at a later date.

The study identified the motivators and barriers to enrolment and retention so that communication and programmatic interventions could be formulated. Some of the favourable conditions that acted as

(Continued)

> **Box 6.3**
> **(Continued)**
>
> motivators were the proximity of the school to all children, availability of basic infrastructure like toilets, number of teachers available in the school, pre-school or early childhood care and education (ECCE) centres, active participation of panchayats and village education committees (VECs), and efforts made by the teacher to ensure enrolment. The barriers to enrolment and retention were the perceived as lack of relevance of available primary education among poor and illiterate parents who were also predominantly from SC/ST communities, the indirect costs of schooling in terms of books and uniforms, opportunity costs of keeping children in school and reluctance of parents to continue sending girls to school after attainment of puberty.
>
> One of the key recommendations of the study was a communication campaign to reinforce the role played by teachers in the education system and thereby enhance their status in the eyes of the community. With the community as the target group, conventional mass media channels of radio and television were recommended. The communication strategy that was developed subsequently built upon the recommendations of the study and suggested a pilot phase in which some ideas could be implemented and tested, as well as a simultaneous consultation with key players in selected states to develop more specific strategies for the states. The development of a symbol for NEEM/UEE and the use of carefully selected excerpts from cinema on issues of primary education and packaging them as TV spots with a message formed the thrust of the communication intervention. Linking the campaign with the 50 years of India's independence then being celebrated across the country with community events to be organised everywhere to take the public discussion on the right to elementary education and UEE forward were other initiatives that were planned.
>
> Unfortunately, the slender majority of the government in office and the lack of consensus on the necessity of introducing the Bill, as well as the uncertainty about the availability of additional resources for UEE, prevented a cohesive approach to the communication effort. Some TV spots were produced and a few of them were aired but there was no campaign execution. This is illustrative of wasted effort, which is often the case with regard to the use of media for social mobilisation. A successful communication requires a concerted effort by various partners as happened in the early years of NLM. Usually, that is not the case.

EVOLUTION OF THE TOTAL LITERACY CAMPAIGNS (TLC)

When it was launched in 1988, the NLM programme strategy was a combination of *a*) the 'centre-based' approach under which a trained animator (or adult education instructor) was paid a small honorarium to run AECs where between 20 and 30 adults gathered for literacy instruction, and *b*) the volunteer-based Mass Programme of Functional Literacy (MPFL) using school and college students on a 'each-one-teach-one' principle. Even though most reviews of these programmes had shown unsatisfactory results, NLM had no other strategy and only hoped to improve performance through better training, monitoring and efficient management.

The main drawbacks of the programme were lack of community involvement and learner motivation, difficulty in monitoring because of the scatter of the AECs in the villages, and the 'each-one-teach-one' scheme not being stimulating enough for the student volunteers. Given its societal goal, NLM had to create a positive environment such that the whole community became geared towards the objective of eradicating illiteracy. For this to happen, the programme strategy had to be supported with a social mobilisation effort. While NLM searched for alternative programme strategies, the first breakthrough came with the success of the Kottayam city literacy effort where the District Collector mobilised 200 volunteers from the Mahatma Gandhi University and linked them to the 2,000 non-literate persons in the city in the age group of 6–60 years and succeeded in making them fully literate in three months (April–June 1989).

However, this was in Kerala, which already had a high literacy rate, and Kottayam city was an urban environment and the numbers to be made literate were therefore limited. Nevertheless, the effectiveness of an intensive campaign in a compact area was noted and it encouraged NLM to support the Ernakulam District Literacy Committee (EDLC) effort initiated in 1989–90. Under the leadership provided by the District Literacy Committee (DLC) that had a few persons from the district administration but also included people from different sectors and varied backgrounds, the goal was to make the district fully literate in one year (Tharakan 1990). The Ernakulam district literacy campaign saw the coming together of the district administration headed by the District Collector, voluntary groups, academics, social activists and business establishments under the umbrella of the EDLS and achieved its objective on 4 February 1990. The success of the Ernakulam campaign lead to the formulation of a statewide TLC in Kerala and the Ernakulam literacy campaign became NLM's programme strategy and 'model' for planning similar district-based TLC elsewhere in the country.

The cardinal principles that emerged from the success of the Ernakulam literacy campaign were:

- A massive and total area approach.
- Involvement of the entire community in one form or the other.
- Predominantly voluntary in nature.
- Shared joy of participation and excitement on achievement.
- Close cooperation of district administration and voluntary workers.
- Very high motivation and productivity of participants.
- Positive change in the social outlook of the participants.

The principal shift in strategy was a change from the scattered and piecemeal approach adopted earlier to focused and well-coordinated, comprehensive and identifiable initiatives. It also marked a change by altering the social context in order to generate demand for the literacy programme and sustaining it. Social mobilisation or environment building, therefore, was a key component of the new NLM strategy of TLCs. Since the TLC depended on a volunteer force to conduct literacy classes, the campaign planning sought to identify learners as well as volunteer teachers ('matching and batching' as it came to be called) through the survey and environment building motivational effort.

The *Kalajatha* and the Importance of Environment Building

One important reason for the failure of earlier (before the 1988 launch of NLM) adult literacy programmes had been the lack of motivation in learners and instructors and an absence of community involvement. The Ernakulam district literacy campaign in 1989 demonstrated that it was possible to enlist community participation and generate a high degree of motivation among learners and volunteer teachers and sustain it through the campaign period. This was achieved through planned environment-building events and activities all over the district so that interest in the programme did not slacken. Everyone—both illiterate persons who wanted to learn and the educated who felt a compelling urge to respond to a social need—wanted to be a part of the literacy programme.

Social mobilisation (or environment creation/building) is an essential component of demand generation on social issues like literacy. It comprises a planned set of activities that achieves multiple objectives. In the literacy campaigns, it was envisaged that social mobilisation would create awareness among all sections of society of the need to become literate; to motivate and enlist a large number of volunteer instructors; to attract illiterate persons and enrol them in literacy classes; to involve various social, cultural and political associations in the literacy effort; to facilitate the formation of community-based 'people's committees' at various levels to implement TLC; and to sustain everyone's involvement in the campaign and prevent dropouts.

Social mobilisation also requires a major communication effort and it is natural that all possible media (multiple media) that are available locally should be used to create the charged atmosphere that is necessary for the success of the literacy campaign. Traditional folk/local media that are available are harnessed for their popularity and their use of the local language and dialect and ability to reach villagers at local fairs and festivals. Other conventional media, including outdoor publicity, print materials and mass media, are also used judiciously in a cost-effective manner. The key element in a social mobilisation effort is detailed planning and coordination that ensures maximum exposure (or publicity) using all the media vehicles.

During the several TLCs that were conducted between 1990 and 1992, certain ground rules for planning environment-building activities were developed that are worth recounting.[16] Print materials were used essentially for the literate audience who had to be motivated to volunteer their time for the literacy campaign and/or support the programme. They had to be kept informed through bulletins, handbills and notices. A few posters were displayed at appropriate places in the district to announce the launch of the literacy campaign or survey, district-level events and completion of the teaching–learning process. Banners and badges were distributed during rallies and conventions to schoolchildren in recognition of their voluntary service.

Merchandising the literacy campaign through the sale of T-shirts, badges and other materials was used to keep up the spirit of the campaign.

Wall writing comprising literacy slogans and the NLM symbol were important means of announcing the presence of the literacy campaign in the district. Hoardings at bus stops and banners in local fairs and festivals, messages in local newspapers and the use of literacy slogans in official communications from the District Collector's office were other important measures adopted to keep the intensity of the TLC alive. Organising visits by the local press and providing human-interest success stories from the TLC to the press and other media were also part of the social mobilisation effort that was crucial for the success of the campaign. All this meant that a dedicated team of volunteers had to be in position to sustain the communication effort. Radio was used during the TLCs to make daily announcements of forthcoming events during the campaign. Radio spots and jingles were used for additional publicity, and discussions were arranged from the local AIR station. Field recordings on the successful ongoing literacy effort in the villages were arranged through visits by the AIR team and hosted by the village-level literacy committees. Doordarshan was used wherever possible, with senior officials participating in discussions. Similarly, wherever available, slides and documentary films were used in cinema theatres so that the district was saturated with media exposure.

However, the unique feature of the TLC was the use of the *kalajatha* as a travelling cultural troupe giving music and theatre performances in villages, motivating the people and enlisting their support towards the literacy campaign. While different districts adopted this form of publicity according to their preference and capability, the general pattern was to use the *kalajatha* in three distinct waves:

1. The *proclamation jatha* that was conducted at the initiation of the TLC, with the *kalajatha* performance in the villages providing an opportunity to enlist volunteers and form village-level committees, motivate illiterate persons and begin the village-level survey work. It was a kind of 'entry point' for the TLC.
2. The *literacy festival jatha* which was more intensive and conducted when the training of volunteer teachers was being conducted. These *jathas* had to be coordinated with other environment-building activities and spread over two to three weeks so that

every village was covered and there was saturation exposure to the literacy messages. This was the most crucial part of the social mobilisation effort and only the well-planned and properly executed calendar of events ensured the massive numbers of volunteers that were necessary for the literacy programme. In some districts the number of volunteers ran into hundreds of thousand during the TLC.
3. The *booster jatha* was undertaken two months after literacy teaching had been started to keep up the spirit of the volunteers and learners. This was a much smaller event to bring the volunteers in the public light and get community appreciation and also push for further enrolment in pockets where it was lagging behind.

The *kalajatha* as a form of rural publicity was popularised by the *Kerala Shastra Sahitya Parishad (KSSP)* as part of their campaign for popularising science in the villages. Using minimum props, simple costumes and little or no make-up, these theatre performances could be mounted at very little cost and relied on community contributions to host the cultural troupe during their visit to a village. Traditional musical instruments and popular folk songs were interspersed with the performance of short plays on themes that were of relevance to the literacy campaign. The success of the *kalajatha* depended a lot on the quality of the scripts, which were usually developed in a workshop with the writer and directors working together to ensure local specificity and entertainment value. Advice from local experts familiar with the traditional/local folk forms of theatre and music was sought to make the performance more attractive and familiar for the village audiences. The several troupes that fanned out to the villages—as many as 20 *jatha* troupes each comprising 10–15 persons performed in the districts during the main literacy *kalajatha*—were trained by master trainers who worked with the scriptwriter, director, music composer and folk media experts. Very often the performers were local students and others with little experience of theatre. Their enthusiasm and commitment to the literacy campaign converted them into literacy activists and *kalajatha* performers. The powerful narratives that touched a chord in the audience and the lively music rooted in the local milieu added to the success of the *kalajatha*s. Held in the evenings, the one or two hour-long performances usually left room for an informal, post-performance discussion with the audience and distribution of publicity materials to interested and motivated individuals.

While the *kalajatha* performance was at the core of the social mobilisation effort, a series of events like rallies and marches, large meetings and conventions were also held where politicians and district officials participated with ordinary literacy workers to achieve the literacy target set for the district as a matter of community pride. It was the social mobilisation achieved through these *jatha*s and supported by the mass media that made the district-based TLC a viable programme strategy for NLM. The following section provides a brief description of the Pudukkottai experience as an example of the kind of planning that went into the social mobilisation effort for a TLC and the spirit of voluntarism that has been the hallmark of a successful literacy campaign.

The Pudukkottai Case Study[17]

The first major effort to mobilise the people of Pudukkottai for the process of change was undertaken at a district convention held on 23 July 1991, in which about 10,000 people participated. A huge procession led by key district officials and the Director General of the National Literacy Mission Authority (NLMA) was taken out through the streets for Pudukkottai town. It was a colourful, enthusiastic and lively event that set the tone for the literacy programme in the district.

The procession started off with the lighting of a torch and the raising of literacy slogans by officials, volunteers and participants. The viewers and passers-by were evidently surprised at the sight of the District Collector, *tahsildar*s and block development officers (BDOs) walking the entire length of the procession and raising literacy slogans. Although at first the officials were a little reluctant to participate in the event, they soon drew inspiration from the volunteers who raised slogans with gusto. The procession stopped at street corners, inviting local people to join in and pledge their mite to the campaign. Public announcements on loudspeakers, beating of drums, music and slogans, and indigenous folk dances added to the festive air in the town.

At the convention an appeal was made to the people of Pudukkottai to volunteer their services as instructors and to help make the 9–45 age group in the district fully literate. Learners were exhorted to enrol in literacy centres as soon as they were opened. A subtle change was being suggested in the relationship between the people and the officials. Till then, the people had sought out the officials for help. The roles had now been reversed: it was the officials and organisers who needed the people's help—as volunteers and instructors.

The success of the district convention set the stage for organising block and village conventions. To a large extent, the district convention was organised 'from above' since the mobilisation was done with the help of the government machinery. The participatory structures had not yet taken root. The block and village conventions were different, as by then a small cadre of *arivoli* (literacy) activists had begun to emerge and the programme was slowly becoming more and more participatory:

> At the Kunnandarkoil and Pudukkottai block conventions where a rally similar to the district convention was held, the local MLAs walked the entire length of the procession and even raised *arivoli* slogans, although they were a little taken aback at being asked to do so at first (Atreya and Chunkath 1996).

Perhaps the most important instrument of motivation in the literacy campaign's first phase was the *kalajatha*. Not many who see a *kalajatha* go back indifferent to the issues presented. And yet, for all its capacity to influence, the *kalajatha* is characterised by its simplicity. The actors wear no make-up and are dressed in simple clothes that are uniform in colour and pattern. There are no elaborate sets; a piece of black cloth and a bamboo stick with a few bells strung on a sash tied at the waist serve as props. These are used to maximum effect, to communicate ideas and meanings. The musical accompaniments are equally simple—the traditional *mridangam* (drum) and a few cymbals can convey the march of a confident woman or the sound of a door opening or the repetitive sound of a typewriter.

Pudukkottai district has 13 blocks and two municipalities. Fifteen troupes were organised so that each troupe could undertake an intensive tour of the block/municipality allotted to it. Each troupe roughly covered 75 villages and a population of about 90,000. The troupes were on the road for about 20 days and gave, in all, about 1,200 performances. On an average, there were around three performances every day; the morning shows were in the local school premises and the evening ones in the village common.

Kaditham (letter) is the story of a poor peasant, Rajamanickam, who was unable to read the letter bearing tidings regarding the death in combat of his son in the army. As Rajamanickam wept his heart out when, at last, the village landlord read the news out to him (but not before he had made him do all the chores at home), many in the audience wept along with him:

> At Kothamangalam village, an old lady clasped the hands of the lead actor and wept because hers had been a similar story. From the tattered ends of her sari, she carefully took out the fifteen paise she had secured there and handed it over to Dharmarajan, asking him to use it for teaching people. It was this kind of generosity that spurred the organisers and volunteers to give their best (Atreya and Chunkath 1996).

Saraswati was a popular play with the women. A young girl dares to step out of her house to study and to learn. She refuses to be deterred by the opposition at home and disapproval in society. Many women in the audience often nodded their heads in approval. Such plays did not evoke much opposition from the men as their services had also been enlisted in the cause of women's literacy.

A visit by the *jatha* troupe helped to generate the nucleus of the village-level committee which shouldered the responsibilities of the programme later on. Food had to be arranged, a stage had to be provided and a public address system had to be installed; for all of these, resources had to be mobilised locally. All these activities helped bring the people together and encouraged them to adopt the programme as their own. The *jatha* would not have been so successful if the artistes had been professionals and if all the arrangements had been paid for by the government.

The voluntary spirit of the troupe members struck a responsive chord in many members of the audience, sometimes in unexpected ways. At Thirumayam village, Kalidas was unable to take part in the day's performance, as he was running a temperature. He had, however, a ready understudy: the van driver, 35-year old Ganesamoorthy who had been taking them round the block. He had been inspired by the troupe members' spirit of service. He ceased to be a hired hand and became one of the *arivoli* troupe, ready to do his bit for the cause of literacy.

There were many sceptics who feared that as Pudukkottai was a poor, backward district, there would not be many volunteers for the *kalajatha* and that it would be impossible to raise local resources for staging the *arivoli* plays and songs. The sceptics were proved wrong and the *kalajatha* enjoyed a rousing reception in almost all the places it visited.

If a group of amateur actors was able to move the people, and make them sit up and take notice, it was because of the passionate involvement of the artistes themselves, because the plays echoed their lives, their fears and their aspirations. Behind the consummate ease with which the troupe members were able to get a sympathetic response from the audience were days of hard work and preparation (Atreya and Chunkath 1996).

THE BHARAT GYAN VIGYAN JATHA (1990 AND 1992)

The appeal of different media is necessarily different. There is no doubt that face-to-face communication is more direct and therefore the participatory mode of cultural performances that are rooted in the specificities of the local milieu have greater attraction and impact. Supported by the interaction through volunteers/activists, this mode of communication had instant, widespread appeal. This was clearly demonstrated by the success of the Kerala TLC programme.

The Bharat Gyan Vigyan Jatha (BGVJ) in 1990 was the first major cultural communication effort initiated by NLM (see Figures 6.6 to 6.8). A large number of exhibitions, processions, songs and performances were organised through a network of over 1,000 mobile cultural troupes in about 350 districts of the country. This kindled the interest of people in literacy and a large number of teachers, students and other professionals came forward to volunteer their time for literacy work. Those who were unlettered, particularly the women, overcame their shyness and reticence and came forward to take part in the literacy campaign.

However, the BGVJ in 1990 did not generate as much enthusiasm as was expected, especially in the north Indian states, mainly because the timing collided with the major political upheavals taking place in the Hindi-speaking states owing to the anti–Mandal Commission demonstrations followed by the *Ram Janmabhoomi Rath Yatra* led by the Bharatiya Janata Party leader Shri L.K. Advani. In view of the limited response it received, NLM decided to launch a second round of the BGVJ in 1992. This was based on the experience in states like

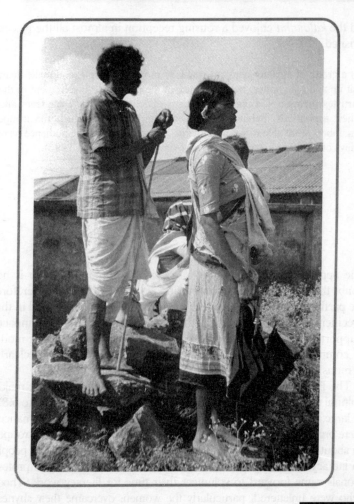

Figure 6.6
A Tribal Couple Watching a *Kalajatha* Performance for Literacy

Andhra Pradesh, Tamil Nadu, Karnataka, Maharashtra and Orissa where the BGVJ had been successful and TLC campaigns had been launched. In order to take the movement further into the Hindi-speaking heartland, a second round of the BGVJ was considered to be necessary and desirable in the low-literacy states. Since the *kalajatha*

Figure 6.7
Young Performers at a *Kalajatha*

Figure 6.8
Village Procession and Rally during the Bharat Gyan Vigyan Jatha

form of using the traditional folk theatre format for social mobilisation on such a large scale had not been attempted before, there was an effort to understand the evolution and role of the *kalajatha* as a communication tool for social mobilisation.

The National Institute of Adult Education (NIAE) followed the planning and execution of the BGVJ in 1992 in selected districts of three states—Rajasthan, Orissa and Bihar—and examined some of the issues thrown up by the use of folk media in a social communication process. These concerned the adaptability of the folk media to the changing socio-economic conditions; problems of incorporating messages of literacy; and the sheer logistics of organising workshops to develop scripts, arranging rehearsals and scheduling the performances in the chosen districts for the travelling troupes. NIAE organised a seminar on 'The Use of Folk Media in Total Literacy Campaigns' in January 1993, where the findings were presented and a broader discussion on the appropriateness of using folk media, the construction of the message and its form, choosing the vocabulary and language of communication, and using folk songs and writing rallying songs or songs of protest took place among scholars, experts and the organisers and activists.

The main findings of the NIAE study were that the volunteers of BGVS were very enthusiastic and committed towards social change and regarded the literacy campaign as an empowering process. In that sense, there was an upsurge of positive feeling that was palpable in the intensity of the preparations—script-writing workshops, play rehearsals and scheduling performances. However, in many places the understanding of the local folk forms was inadequate and sometimes the suggested themes or ideas given to the local artists were too rigid and did not allow for flexibility. For instance, in one play a person dies and is brought before Yama (the god of death). He pleads with Yama to give him some more time on earth to become literate. His wish is granted but he whiles away the time and dies and is taken to Yama again. Again, he makes the same plea and returns to earth. This happens three times and finally the person learns to read and write at the end of the play. The treatment was light and humorous and the local form was used appropriately to communicate to adult non-literates that they should use the present time to become literate. However, the organisers felt that 'obscurantist' ideas like Yama should not be used in the context of promoting a scientific temper through the literacy movement, and the particular play was never performed.[18]

In several districts, a large number of the volunteers were schoolgoing and college students who were motivated and enthusiastic about doing something. The performance design, therefore, was somewhat urban and followed the 'street theatre' format of short-duration plays staged in urban settings. The language also used a more urban, literary form rather than drawing on the rich oral tradition and culture of the rural areas. While the subject was very topical, the short and intense street theatre style of treatment, which is appropriate in cramped urban neighbourhood setting, failed to make the same impact in an open rural settings. A case in point was a play on *Mrityubhoj*, a practice in rural Rajasthan where poor families have to arrange a lavish feast upon the death of the head of the family, which very often led to further indebtedness. While the play, which questioned this practice, was powerful and very relevant, it did not hold the attention of the audience as its form was too taut for an audience accustomed to a more leisured form of communication that touched their emotional chord gently and more easily.[19]

The issue of folk media as an appropriate tool of communication raised several questions, especially in the context of a government-sponsored initiative like the NLM. Historically, theatre and songs had been used as a form of social communication for a long time in India and during the Independence struggle the popularity of some of the plays of the Indian People's Theatre Association (IPTA) caused considerable worry to the government in power. Subsequently, various political groups have used theatre, in both urban and rural settings, to make people aware of issues by requiring them to critically reflect on their social reality and questioning existing social relationships. This followed the pattern of Brechtian theatre of social commitment. The success of Utpal Dutt in West Bengal, Gursharan Singh in Punjab, Gadar in Andhra Pradesh, Safdar Hashmi's Jana Natya Manch in Delhi, Prasanna's Samudaya in Karnataka and KSSP's effort in Kerala, are all examples of successful endeavours using theatre as a mode of popular communication to arouse social consciousness and bring about change. The main point to be noted was that these performances were largely supported by community contribution and people's voluntary support. In the course of time several NGOs and voluntary agencies adopted the same format to create awareness on several development issues. The BGVJ was a very large-scale effort of the NLM to amalgamate the value of critically questioning existing social conditions (including the oppressive structure of the State and its agencies) and promote the literacy programme of the government as an empowering tool for the poor. Following the theory and practice of Paulo Freire, adult literacy provided the right platform to take this initiative.

The question was: would it be possible to be critical of the government and the social structure while receiving funds from the government for the *kalajatha* performances? The BGVJ was a fairly successful demonstration of the fine balance that had to be maintained between the compulsions of the political activist using theatre (*kalajatha*) as a tool to arouse social conscience against injustice and oppression and, at the same time, promote a government programme without losing credibility or compromising its principles. Fortunately, NLM was fairly broadminded and recognised that adult literacy had to have an empowering content and allowed more than a fair degree of freedom to the BGVJ organisers. On the other hand, the BGVJ organisers sometimes failed to respect the views and sensibilities of the audience in their eagerness to make their point. This often created a gap between the audience of stratified village folk and 'outsiders' presenting the village reality and exhorting individuals towards action and social change.

The NIAE study and deliberations at the national seminar showed that the BGVS grappled with all these issues in the course of their long journey of organising thousands of *kalajatha* performances spread over hundreds of districts all across the country. Not everywhere did the performers meet with the same degree of enthusiasm or response from the community as they did in Pudukkottai. In several places there was a mature team that worked well and the success of the *kalajatha*s (BGVJ) was measured by the prompt initiative by the local community leadership to take up a TLC. In other places, the performances touched the lives of the poor and they connected with the performers and were motivated, but there was inadequate follow-up and the literacy movement could not convert their tentative enthusiasm into enrolment in literacy classes. In conclusion, it must be said that despite the lapses and failures, the 1990 and 1992 BGVJs were successful social mobilisation efforts for adult literacy and basic education. They were also on a scale that had not been attempted before. Subsequently, *kalajatha*s and environment building became an integral part of the planning format of TLCs but failed to ignite the spark created by the BGVJ for literacy. In the following section we shall take a brief look at the limitations of the TLC approach and the reasons for the gradual petering out of the National Literacy Mission.

TOTAL LITERACY CAMPAIGNS—THE AFTERMATH

The achievement of the Ernakulam District Literacy Campaign and the subsequent social mobilisation through the BGVJs in 1990 and 1992 had made TLC a programme strategy for NLM to achieve its target. The early phase of the TLC approach (1990–92) was an exhilarating experience for those participating in this massive effort. The coordination between the NLM authorities in Delhi and the Zila Saksharta Samitis (ZSSs) in the districts implementing TLC with support from the state governments was something new for a government programme. The involvement and support of non-governmental organisations like the BGVS at all levels to facilitate the implementation of TLC added to the hectic but smooth expansion of the TLC programme. Adequate planning and preparatory work was done prior to the launch of a TLC in a district to ensure that a core team of committed individuals was available at the district to take charge of the time-bound execution of the campaign. The design of the TLC was also left flexible to incorporate the particularities of the local context.

The expansion of the TLC programme was consequently cautious and slow, building upon the strength of the success of earlier TLCs. It is also noteworthy that the initial success of the TLCs came from the relatively more progressive states with a history of social work and voluntary action. The more difficult Hindi-speaking heartland had not been tackled successfully. Unfortunately, the success of the TLC effort resulted in allocation of much greater funds to upscale the programme to cover a large number of districts quickly. The aftermath of the destruction of the Babri Masjid in 1992 and the consequent communal tension was also a cause for concern, particularly in the Hindi-speaking states. The TLC, particularly the social mobilisation effort, was perceived as an opportunity to inculcate values of communal harmony and national unity. Many TLCs in several districts of the Hindi-speaking states were launched without adequate planning and preparation and funds were disbursed quickly. Social mobilisation in terms of *kalajatha*s and other rallies and marches became a routine formula without the commitment and enthusiasm of the earlier volunteers. The scale of each campaign with lakhs of persons to be made literate in each district became a daunting target for even the most enthusiastic volunteers.[20]

The districts that had done well in the early phase of the TLC and had succeeded in making a large number of adults literate (achieving a fragile level of literacy) did not have a post-literacy and continuing education programme to reinforce and firm up the literacy achievement. A few districts managed to cope with the pressure of sustaining the pace through voluntary effort, but after a while fatigue began to show even in these districts.

NLM's advertising campaign also changed course and the old theme of *Chalo Padhayen, Kuch Kar Dikhayen* was changed to an appeal to join the largest 'civil or military' mobilisation of people since the freedom movement. The tenor of the campaign was altered and the leadership for implementing the TLC fell on the District Collector's lap with little support from the NGOs. Consequently, teachers and other government officials were pressured to enlist as supervisors in the TLC. The earlier volunteer spirit gave way to indifference and cynicism among teachers and government servants forced to take up literacy work. The major task was to somehow spend the large sums of monies that had been released to the districts by NLM—a task which became difficult in the absence of social mobilisation and the enlisting of volunteers and learners. The lack of a comprehensive and sustained continuing education programme left the neo-literates without support and relapse into illiteracy became a real problem. In addition, NLM's messages lacked credibility. The whole programme thus lost its momentum and petered out before the turn of the century.

Rural Newspapers—A New Media Vehicle for Neoliterates

After the initial spurt of successful TLCs in the early 1990s, it was recognised that a large number of persons had become literate though their achievement was fragile. They needed to practice their recently acquired literacy skills regularly. In rural areas there is hardly any challenging enough reading matter readily available for the neoliterates to practice their literacy skill. Penetration of newspapers into rural areas was low and other printed matter that was in circulation (wall writings, hoardings, film posters, leaflets, etc.) were quite limited.

The NIAE carried out an experiment to determine the form and content of a media vehicle that would be appropriate for neoliterate persons since the regular newspaper was too dense for them to read easily.

A weekly broadsheet was designed and tested among the neoliterates for technical specifications like point size, sentence length and word length of news items, reports and stories (see Figure 6.9). When the prototypes in Hindi, Tamil and Bengali were field-tested, the response of the neoliterates was positive and indicated that such weekly broadsheets would be received well and have a readership in the village.

NIAE also conducted a study of the newspaper outreach/distribution system at the village level and found that there was no mechanism to reach newspapers to the villages. Newsagents and hawkers, even of district papers, were restricted to the towns and surrounding areas. Local newspapers operated on small margins and did not have the motivation or capacity to create a distribution network. On the other hand, the NIAE study showed that a significant percentage of neoliterate consumers in rural areas owned key assets like housing and durables like wristwatches, radios and bicycles. Similarly, there was high usage of consumer goods like washing and bathing soap, hair oil, batteries and talcum powder. This indicated that neoliterate persons in rural areas represented a significant market segment and that it might be possible to get advertising support for a rural newspaper.

NIAE shared its findings at a national seminar in 1992 and a large number of ZSS and representatives of BGVS were enthusiastic about the possibility of having a rural newspaper. With the support of funds made available by NLM under the post-literacy programme, some of the ZSS started a weekly broadsheet for neoliterates. The most successful and noteworthy was *Velugu Bata* brought out from the Chittoor district, which also demonstrated the potential of desktop publishing since the entire editorial, design, production and distribution was handled by one person with some assistance from others and the entire operation was carried out in Chowde Palle, a village near Chittoor. *Velugu Bata* made history when it was forced by the state authorities to suspend publication for a brief while during the peak of the anti-arrack movement in Andhra Pradesh. This anti-arrack movement started in the neighbouring Nellore district where a group of neoliterate women read a lesson in their post-literacy primer about a woman who had led a movement to stop the sale of arrack in her village by mobilising the other women of the village. Through the network of literacy volunteers the movement spread all over Nellore district. *Velugu Bata* covered the story in its weekly broadsheet and the state authorities felt threatened since the anti-arrack movement

Figure 6.9
Broadsheets in Different Languages for Neoliterate Persons

was already snowballing into a major political crisis for the state government. The district authorities were put under pressure and *Velugu Bata* had to stop publication for a while. This incident not only demonstrated the potential power of the press and the perceived fear among officials, but also pointed to the importance of making appropriate information and reading materials available to neoliterates.

Some districts of Tamil Nadu priced their broadsheets and organised reading groups in villages (Figure 6.10), while other districts succeeded in getting modest advertising support. However, these efforts could not become self-sustaining. In Banda district of Uttar Pradesh, an initiative started by NIAE (followed through by Nirantar, a women's resource centre) took root and a weekly newspaper—*Mahila Dakiya* (later called *Khabar Lahariyan*)—developed in collaboration with Mahila Samakhya. This newspaper, which allows the women themselves to write and design their own materials, is flourishing with a monthly circulation of over 10,000 copies. Nirantar also provides a feature service of appropriate reading materials for neoliterates, which many NGOs and other voluntary groups and ZSS pick up and use in their post-literacy and continuing education centres. However, these efforts have largely depended on grants from NLM or some other

Figure 6.10
Women Learners Reading Lessons

funding agency. The mainstream newspaper industry has not been attracted to venture into the world of neoliterates to cultivate a new readership.[21]

Aajkaal, a Bengali newspaper of the *Ananda Bazar Patrika* group, was enthused with the idea and planned to bring out a four-page tabloid newspaper in Bengali tailored for the neoliterate and make it available to them free of cost till they reached a stable state of literacy. They wanted to try it out in one district—Burdwan—and initially restrict the distribution to the non-municipal areas to test the concept. With 30,000–35,000 copies to distribute, *Aajkaal* hoped to reach most of the 300,000 neoliterates in the approximately 3,000 villages in the district. It hoped to use the literacy volunteers as the agents for distribution and, as an incentive, offered them the Sunday edition of the mainline paper with its colour supplement free of cost. *Aajkaal* hoped to cover a major share of its costs through advertising support though it felt that it would still need a small subsidy to make the project viable. While UNICEF was willing to provide the requisite subsidy amount for one year, the entire process took so long that the commercial owners of *Aajkaal* lost interest and the project was abandoned after three years of discussion and experiment with format, content, advertising and distribution mechanisms.

While the *Aajkaal* attempt was innovative and important, it also demonstrated the fact the mainstream newspaper industry was not convinced of NLM's claim of having made a large number of persons literate in the rural areas. It did not see sufficient potential of expanding their readership base in rural areas in order to make the necessary additional investment. The fact that television coverage was expanding much faster than penetration of newspapers in rural areas showed that spread of literacy and inculcation of the reading habit has not yet registered as a market demand for reading materials in rural areas. NLM tried to provide a fillip but did not pursue the private sector newspaper industry sufficiently to enlist their participation and partnership. *Aajkaal* was the sole exception that did not quite take off.

NOTES

1. A publication titled *Challenge of Education: A Policy Perspective* that encapsulated the issues raised in the discussions with a wide cross-section of public opinion was brought out by the Ministry of Education, Government of India in August 1985. Many of these issues found place in the New Policy on Education 1986 issued by the Government of India.
2. The meeting of heads of governments of most countries in Jomtien, Thailand, in 1990 under the aegis of the United Nations was a major initiative of UN agencies and international donors to get the various governments' commitment to address issues of basic education facing the developing nations and also persuade the developed countries to help meet the gap in financial resources. The Education for All (EFA) Declaration adopted in Jomtien to achieve Universalisation of Elementary Education (UEE) by 2000 became the target to be reached through development aid and programmes for accelerated literacy and basic education.
3. Various kinds of cultural *jatha*s or mobile performances were organised in districts during the NLM's literacy campaigns and contributed to the success of the Total Literacy Campaigns (TLC). See later in the same section for more details.
4. The National Literacy Mission (NLM) document published in 1988 at the time of the launch of the NLM set down these objectives in clear terms.
5. Jean Drèze and Amartya Sen have persuasively argued the importance of the 'agency of women' in development and made a case for greater investments in their health care, basic education and opportunities for employment.

6. NLM's communication strategy was developed through a series of meetings and discussions with communication experts and a workshop held in Mysore in 1988 with practising adult education functionaries from the State Directorates of Adult Education (SDAEs) and State Resource Centres (SRCs).
7. A very articulate and succinct statement of this position is presented in a short mimeographed paper by Kamala Bhasin (1984).
8. Two series of TV films (*Jaag Sakhi* and *Khilti Kaliyan*) were produced by the DAE in the period 1985–87 prior to the launch of the NLM. These films, which were produced with the support of UNICEF and promoted the cause of women's literacy, were part of a range of films produced by DAE for motivating adult learners through film screenings in villages and through television.
9. The study was later published by UNICEF under the title '*The Woman as Learner—Not Just Another Statistic*' (see Kapoor and Unnikrishnan 1990).
10. Safdar Hashmi was a very well-known name in theatre, particularly to those involved in using street-theatre as a tool for political activism. However, Safdar's creative work in literacy is less known.
11. I was directly responsible for the PREAL project acting as its Co-ordinator in DAE. A narrative account of the project was submitted to UNICEF in 1991. A meeting to review the project was also held under the auspices of DAE. The present account draws on the discussions at the review meeting and my own experiences during PREAL.
12. The feedback study was conducted by DECU, ISRO, and submitted to DAE, New Delhi in 1991.
13. A film titled *Beyond the 40th Episode* was produced by DAE to document this action-research experiment and study conducted by State Resource Centre for Adult Education, Jamia Millia Islamia, New Delhi, for DAE and UNICEF.
14. Roda Mehta, who was then Research Director in Ogilvy & Mather, gave an account of this strategy and NLM experience in a two-part article in *Brand Equity* (*The Economic Times*), 19 and 26 February, 1992.
15. The production of TV/radio spots and a mass media strategy to enlist volunteers for literacy was based on the communication strategy, including a Creative Strategy and Mass Media Plan for NLM (1989) developed by Ogilvy & Mather Advertising for DAE and accepted by the NLM. The theme 'Mission for the Literate' was given by the agency to focus attention on the importance of educated volunteers in NLM's literacy effort.
16. As the TLC movement spread to many districts, the Bharat Gyan Vigyan Samiti (BGVS), which was instrumental in forging the link between the district administration, academic community and volunteers, developed

a guide book for a mass campaign for total literacy to set down the different stages of the planning and implementation of a TLC.
17. See also Atreya and Chunkath (1996) for a more detailed account of the Pudukkottai TLC.
18. Case study of the Orissa BGVJ 1992 presented by Runu Chakrabarty at the NIAE seminar on the Use of Folk Media in Literacy Campaigns in January 1993 (mimeographed).
19. Case study of the Rajasthan BGVJ 1992 presented by Aisharya Kumar at the NIAE seminar on Use of Folk Media in Literacy Campaigns in January 1993 (mimeographed).
20. A more detailed objective analysis of the TLC is available in the Report of the Expert Group appointed by the MHRD, Government of India to evaluate the literacy campaigns in India.
21. For more details on the post-literacy efforts and preparation of reading materials for neoliterate persons, see also the 2002 report of Bharat Gyan Vigyan Samiti on the Impact of Post Literacy in India, a study sponsored by UNESCO.

Population: Bringing about Behaviour Change

India was one of the first countries to adopt a family planning programme on a national scale. The First Five-Year Plan (1951–56) declared that the 'higher the rate of increase of population the larger is likely to be the effort needed to raise per capita living standards'. The government has over the years sponsored various measures aimed at encouraging limitation to birth and the communication media have played an important role in this massive effort. In 1966, the Mass Education and Media (MEM) organisation was created within the Department of Family Planning to promote the small family norm. A network of field personnel at the national, state, district and block levels were engaged, following the pattern of the agricultural extension services, to reach out to people and inform them about contraception and motivate them towards family planning. Radio and film also began to be used in a limited way. It was in this period that a strategy for communication and a pinpointed, clear and specific message to the family were articulated for the first time. The *Red Triangle* symbol for family planning was developed and slogans propagating 'two or three children—enough' and the small family norm started being used as a mass campaign.[1]

In 1970 the campaign for commercial distribution and sale of Nirodh (condoms) through the retail outlets of a few major consumer goods companies was launched. This was done after considerable research (studies on Knowledge, Attitudes and Practices [KAP]) and with the support of professional advertising agencies. The THINK campaign, as it came to be called, promoted the concept of spacing and supported the marketing of the Nirodh. Owing to their powerful impact as an audiovisual medium, films were also seen as a major vehicle of communication and the district units of the MEM were equipped with audiovisual vans for exhibiting these motivational films.

At that time, the current perception was that the problem was one of dissemination of information and motivation of potential acceptors of the family planning programme. Hence, the communication effort was geared to wide publicity through the use of mass media and interpersonal communication (IPC) through extension agents. The use of mass media through the centralised structures of broadcasting and film production matched the approach of centralised planning that relied on the expertise of urban planners. However, the failure of development programmes to mitigate the conditions of extreme poverty raised questions about the centralised approach to family planning communication. It was argued that the participation of people in the planning process as well as in execution was essential. The extension approach, highly successful in European and other contexts, did not prove appropriate in the Indian cultural context. The centrally produced films, dubbed in different languages but ignoring the regional variations in language and customs, were rendered meaningless.

A review of the family planning communication programme in 1973 identified several factors as being responsible for 'resistances to family planning'. Some of these were the need for additional helping hands for poor farming families to supplement family income, fear of infant/child mortality and preference for sons for old-age security.

Unfortunately, during the Emergency period (1975–76), the combined force of mass sterilisation camps (with some element of coercion) and the imposition of press censorship that suppressed negative feedback, created a horror of family planning in the public mind. While targets may have been achieved in that period, the use of force, as has been practised in many countries to bring down population growth rate, could not be a path that India could follow and planning one's family had to remain the voluntary choice of married couples. The planning of the communication effort thus had to be more sensitive to needs assessment and socio-cultural specificities of particular audience segments. Reliance on mass media had to be tempered with adequate emphasis on interpersonal communication (Banerjee 1979).

While the family planning communication programme in the 1960s and 1970s established that creating awareness and generating demand were possible, it also showed that supply of services could not be restricted only to contraceptives and sterilisation but had to include public health facilities and care for the mother and child. It showed that resistances have to be addressed and in a democratic polity, the poor and powerless cannot be presumed upon. Much research work and field studies were executed during the early period of family planning campaigns. Time and again, it was pointed out that the reach of mass media was limited and that IPC channels had to be used through the functionaries and other opinion leaders. It was also evident that the communication effort succeeded better where there was clear evidence of higher social development and political backing to the programme.

By the turn of the 1980s, coinciding with the Sixth Five Year Plan (1979–84), the programme had been re-designated as the Family Welfare Programme (FWP). The focus shifted from sterilisation and contraception to concerns of child survival and maternal health, with the adoption of family planning methods left as a matter of voluntary choice. At least policy-makers began to talk of the need to elicit people's participation in the planning process and Information, Education and Communication (IEC) was perceived as a valid activity in the planning and implementation of development programmes. Another important element for communication was the focus on barriers/resistances. Removing the bias against the girl child, making people aware of the legal minimum age for marriage of girls (18 years) and improving the status of women in the family and society were some of the messages initiated in this period. It was also a period when the importance of IPC was recognised and a countrywide training programme in interpersonal communication skills for the grassroots-level functionaries begun. Population Education in the school curricula and through the adult education centres was also introduced.

The lure of commercial, colour television and consumerist culture began in the mid-1980s and the concern for family welfare (safe motherhood and child survival) and population control was expressed through numerous TV spots and quickies that were aired daily. As audiences were being exposed to the popular Hindi cinema through television, communication for family planning also became more open with discussions on reproductive physiology and contraception becoming more acceptable even on television for family viewing.

Hum Log, a television serial that used the 'soap opera' format, focused on the social theme of women's status and family planning/welfare.

The inspiration for the serial came from the work of Miguel Sabeido in Mexico, and it was quite popular among urban/semi-urban middle-class audiences (being the first TV serial to be telecast by Doordarshan). However, the messages were ineffective as the viewers had already achieved a level of awareness through a process of education and economic betterment. This experience showed the inability of our communication planners to effectively adapt an idea with the research rigour that was required. Popularity as a social drama on TV did not justify the investment in terms of the initial social objectives.

Reviewing the Seventh Plan period, the Eighth Plan (Government of India 1992–97) clearly mentioned that IEC was weak and that the focus was still on national issues rather than personal issues for choice by couples. Stating that the 'incongruity of perception between the people and providers of services' had cost the programme dearly, the document articulated the need for a National Population Policy (NPP) and argued for a convergence of services. According to the Eighth Plan, 'Based on a holistic approach to social development and population control, integrated programmes for raising female literacy, female employment, status of women, nutrition and reduction of infant mortality will be implemented.' Emphasis was laid on decentralised planning and implementation with an increased role for Panchayati Raj Institutions (PRIs) as well as voluntary agencies building up to a 'people's movement with total and committed involvement of community leaders and linking population control with female literacy, women's employment, social security, access to health services and mother and child care'. The Plan document also stated that the 'backbone of the IEC effort will however remain the interpersonal communication for which the grassroots level female worker will have to be trained and effectively utilised'.

IEC STRATEGY DEVELOPMENT FOR FAMILY WELFARE

In order to achieve GOI's goals of family welfare by 2000, an action-oriented agenda was set down that comprised expanding access to services, improving their quality and increasing the demand for them. In this connection, the Ministry of Health and Family Welfare

Table 7.2
Demographic Issues

Age at marriage	▸ Female education must be an essential component of IEC strategy.
	▸ Removal of social stigma attached to late marriage may require more urgent and intensive focus than 'too young to bear responsibility' approach.
	▸ Communication messages pertaining to age of marriage for boys should incorporate aspects of working status linked to appropriate time to marry.
	▸ Girls do not possess any decision-making power at the household level with respect to their marriage. A long-term communication strategy may aim at addressing this issue of age at marriage more intensively to the elders and the men.
Family size and composition	▸ The requirement of minimum one son is so ingrained that no short or simple IEC device can be expected to counter the same in the short run.
	▸ Three target segments for communication in this regard—adolescent girls, men and mothers-in-law—should be given individual and separate treatment.
	▸ There is an urgent need to seek repositioning strategies on this which are novel yet address the basic issues.
Adoption of FP methods—techniques	▸ Temporary methods of birth control should receive higher emphasis in FP communication.
	▸ Communication strategy must not only adopt a proper media mix but also put a greater emphasis on interpersonal methods. Small group meetings in settlements encouraging discussions and question-answer sessions will help dispel fears and misconceptions.
	▸ Senior officials including doctors should also be involved in interpersonal communication in group or village settings.
Adoption of FP methods—Spacing	▸ Spacing must feature as the most important FP technique for adoption if the operational objective is to be successfully shifted to 'demand generation' rather than target fulfilment.
	▸ While mass media must be used for awareness generation, a very strong emphasis on interpersonal communication at the local level must be stressed.
	▸ Adolescent girls, especially those about to be married, and newly married couples must constitute distinct and important target segments for communication.

Source: ORG 1992.

spacing. A brief description of these specific recommendations are given in Tables 7.1 and 7.2. The aspects highlighted for message and media were message content and execution, mass media, traditional/folk media and interpersonal communication.

Message and Media

Message Content and Execution

- Message content must treat the target segment as rational decision-makers and not gloss over negative aspects.
- Message development for interpersonal communication must be taken very seriously.
- Nature and content of messages suitable for different media should be ascertained and subjects closer to the home environment of the target segment should be woven into the messages.
- The concept of 'enter–educate' must take firm root in FW communications.
- Even though strong son preference is the prime determinant of family size for economic considerations, the emotive aspects of having children, whether boys or girls, should also form 'an integral and important part' of message development.
- Planning for communication campaigns must follow a multi-stage approach.
- Research must form an integral part of message development, and activities such as audience research, pilot testing of materials and field validation of the messages must take place.

Mass Media

- For each of the two major media—TV and radio—the content and style of presentation should be properly assessed. TV has a very high potential for 'awareness generation', which must be exploited. Radio should also be used innovatively such as for story-telling in the folk/traditional style, which is very popular in rural India. Local TV channels must be actively involved in FW communication.
- The reach of wall paintings/hoardings should be exploited. Hoardings are one of the most effective ways of communicating

to new and illiterate populations. The visual literacy of understanding symbols and expressions displayed, along with the long life and continuous reminder characteristics of the medium, should be used as an asset. Creatively adopted, hoardings can be used even to tell a story very effectively.

- Limitations of these media must be taken into account and the IEC strategy should use a judicious media mix rather than become over-dependent on mass media. While public awareness, reinforcement of messages, and public recognition can be addressed by these media, behavioural change can only be effected when IPC is properly used.

Traditional/Folk Media

- Traditional/folk media must be immediately brought within the ambit of communication planning and used in a focused manner for selective subject communication. For example, awareness generation should be left largely to the mass media.
- New administrative procedures in the IEC planning and delivery structure will need to be introduced as the use of folk media will entail flexibility as well as mutability of the message content. This is not possible under the current system.
- Communication strategies should be explored in association with other kinds of agencies. For example, in selected pockets, NGOs can be entrusted to fully supervise the folk media campaign.

Interpersonal Communication

Interpersonal communication must be taken as the most important segment of the IEC strategy:

- Upgrading skills and improving the morale of frontline workers were important elements that needed addressing. To do this would require a heavier emphasis on training as compared to its current status. Skills to improve group communication as distinct from face-to-face communication needed to be imparted.
- The morale of workers was a complex issue that needed to be addressed urgently. It would therefore be necessary to look at alternative options for evaluation of performance. Feedback

from frontline workers could provide invaluable inputs to communication campaigns—for example, their opinion on 'how best to communicate' could provide valuable insights for the formulation of a communication strategy. Involvement of frontline workers in the feedback mechanisms would therefore be a must.
- Morale could also be addressed by regular supply of communication equipment and aids.

This study by ORG/JHU and the IEC strategy recommendations were not implemented by the MOHFW, though some of them like increasing reliance on IPC and traditional/folk media gained ascendance in subsequent IEC plans. The 1994 International Conference on Population and Development (ICPD) in Cairo marked a major policy shift with its emphasis on gender equity and reproductive health and a move away from the operational objective of population control. While the Indian FWP was evolving towards the 'new paradigm' with a target-free approach (TFA) and focus on reproductive and child health (RCH), the United Nations Population Fund (UNFPA) took the initiative to develop an IEC strategy to support the Government of India's effort to widely disseminate the recommendations of the ICPD.

ICPD AND AFTER: PARADIGM SHIFT

The effort of UNFPA was to further the recommendations of the ICPD held in Cairo in 1994, particularly those linking population issues with sustainable development and gender equality, thus broadening the agenda of MOHFW's family welfare programme. It was, however, felt that some data on the perceptions of people on issues of population and development, including their views on gender and family welfare, would be useful in formulating an advocacy and communications programme. Such a study would also provide some insights into the changes that may have come about in perceptions, if not in practice, over the years.

Accordingly, a study was assigned by UNFPA to Mode Research to ascertain the views of people and thereby provide information that would help formulate the IEC strategy for UNFPA's Country Programme. The study[3] was conducted in four states—Tamil Nadu,

West Bengal, Rajasthan and Madhya Pradesh. An exploratory phase comprising discussions with experts and a qualitative study using focus group discussions and in-depth interviews contributed to the final design of the research instruments and preceded the execution of the survey. The broad conclusions that had significance for the development of an IEC strategy were:

- A clear majority of households in the sample were nuclear families.
- Population as an issue did not find salient mention.
- Burning issues were unemployment, rising prices, housing and availability of water.
- However, on goading, the linkage between increasing population and rising unemployment was admitted.
- Inaction by government perceived as major cause for the ills of unemployment, etc.
- Importance of people's participation in development programmes not given importance except in West Bengal.
- Improvement in living conditions over time was perceived but standard of living was regarded as the same.
- Significantly, improvement in health service delivery was not perceived as a sign of better times.
- Erosion of values noticed—more money-minded, more self-centred and less sharing.
- Heartening feature was improvement in women's condition—less purdah, more respect and greater mobility.
- Better communication between husband and wife admitted as a sign of better times.
- Small family norm and delayed age of marriage were accepted. In Tamil Nadu it was a reality as well.
- Clear bias for boys continued. Resistance to delaying first child after marriage was very strong.
- Importance of girls' education was recognised. Education as route to empowerment and independence was admitted.
- Yet, eventually, marriage and childbearing were seen as only role for women. Their productive role went unnoticed.
- Awareness of FP methods was there. Competent knowledge of spacing methods was lacking.
- Contact with field functionaries was limited. Functionaries were poor in IPC.

- Exposure to mass media was limited, especially among women in Rajasthan. IPC was very necessary.
- Functionaries did not regard IPC as their job. Most were not trained in IPC.

It was significant that resistance to delaying the first child after marriage remained strong despite the sustained motivational/communications campaign in this regard. Social pressures and the fear of being marked as infertile were some of the reasons given by the respondents. On the other hand, women and men were receptive to the idea of temporary methods of contraception and interested in spacing, but only after the birth of the first child. There was therefore demand for temporary methods of contraception. The low contraception prevalence rate was probably because of insufficient promotion and availability and supply. The thrust of the FWP was undeniably towards terminal methods and that too for female sterilisation.

The perception that girls should get married at an older age was accepted and most respondents indicated that prevalent practice was for girls to marry at 17–18 years. Education for girls was universally accepted and believed to be necessary by the respondents. However, that did not deter them from confirming that the most important thing in life for a girl was to get married. Nor did it shake the preference of respondents for male children.

If we look at these perceptions, it would appear that there could be two approaches to tackle the resistance. Accept it and continue with the arduous task of overcoming resistance to delaying the first child after marriage and/or weaken the preference for sons. Alternatively, a different route building on the positive changes that have taken place could be tried. For instance, if the age of marriage at 18 years had become acceptable (more or less) and the importance of educating a girl was recognised, perhaps there was scope to argue that it is necessary to delay the age at marriage even further to 20–21 years so that the girl has matured completely and also had the opportunity for education and other skill development to improve her earning capacities.

Similarly, the survey showed that there was perceived improvement in communication between husband and wife and that they appeared to be sharing the parenting burden at least in times of illness and when visits to the doctor were required. Perhaps this had become necessary in the context of nuclear families, which appeared as a majority in the sample survey. In this context, a communications effort that acknowledged this change and built on it to suggest greater male responsibility in sexual/reproductive behaviour and parenting, reflected through greater respect and a more caring attitude towards the wife, might be effective. It would lead not only to lightening women's burden in the family, but also encourage them to demand better reproductive health services.

Priority Communication Tasks

It was evident that advocacy, whether for gender equality and sustainable development, reproductive health services, male responsibilities, involvement of NGOs or population education for adolescents, was a high priority. Whether these issues should be addressed through a national media campaign or on a more modest scale might have been a debatable point. However, what was also apparent was that the focus of the communications project needed to be in those particular states and districts where the FWP had lagged behind.

Gender equality and equity was a fundamental concern expressed in the ICPD Programme of Action (POA) and this could be seen both in the emphasis on reproductive health as well as in the focus on sustainable development. Advocacy for gender equality and equity was necessary since the attitude of the programme planners needed to be altered from the prevalent perception that women, particularly poor women, were responsible for overpopulation. At the same time, it was realised that changing attitudes could take a long while, particularly in a society where male domination is so strongly manifest. Action would have to be initiated through appropriate projects where some of the concerns expressed for reproductive health services might be demonstrated through better-designed health service delivery. Communications support would be crucial to prepare the ground prior to commencement and during the execution of such projects.

The complementary aspect of gender equality and equity was increased attention towards male responsibilities both in terms of reproductive behaviour and parenting. It was significant that the Mode Research study showed that in both urban and rural areas, nuclear families tended to dominate. Coupled with the finding that women admitted better communication with their husbands and saw this as a sign of positive change from earlier practice, the opening for a communications effort was already there for altering male attitudes. There was also a need for information/education on male and female reproductive physiology and competent knowledge on spacing methods. The finding that husbands do lend their support for childcare in times of illness could become a positive entry point for urging men to do a little more to lighten the burden on women.

The study also showed very clearly that contact with field-level functionaries was very limited and adequate information about family planning, let alone reproductive health, was lacking. This was largely because the functionaries did not perceive IPC as being part of their job. Very often, they were not trained in IPC skills. While it was undeniable that it was necessary for functionaries responsible for the delivery of services to be trained in IPC, it was also time to think of other channels as well. It is in this context that the role of panchayats and grassroots-level voluntary agencies/NGOs required serious consideration. Being closer to the community and having regular contact with them, panchayats and grassroots-level voluntary agencies/NGOs could provide crucial support to the IEC effort to promote reproductive health (RH) services through their IPC with the community.

With the adoption of a 'target-free approach' and introduction of the decentralised, participatory planning process based on community needs assessment, it appeared that the operational objective of the FWP had finally made a definite break from its past preoccupation with meeting sterilisation targets. The countrywide Reproductive and Child Health (RCH) Project introduced in October 1997 covered an expanded range of services to address the hitherto unattended issues of unwanted fertility, maternal health and child health.

Moving from Family Welfare to Reproductive Health: Implications for IEC

The essential RH package necessitated a range of communication activities that were broader in scope than before. In addition to prevention of unwanted pregnancies and the promotion of childhood immunisation, issues of safe abortion, medical termination of pregnancy (MTP), safe motherhood (prevention of maternal and infant mortality), prevention and management of Reproductive Tract Infections/Sexually Transmitted Diseases (RTIs/STDs), child survival, as well as health, sexuality and gender had to be addressed.

These goals required a strategic approach that identified meaningful segments of the target audience, promoted new behaviours that were closely linked but varied, identified messages that reflected the perceived benefits to be gained from each of these behaviours, and used a mix of communication channels to reach various audiences. It was also necessary to identify barriers—both external (lack of access to services) and internal (misinformation or fear of side effects)—to the adoption of new behaviours and develop strategies to deal with them.

In short, the following three needs were to be addressed immediately:

- The need to devise a well-defined, area-specific, focused communication strategy.
- The need to decentralise the planning process to the district level, which included building the FW/RCH programme's capacity to communicate with the target audience.
- The need to track changes in knowledge, attitudes, beliefs and practices to provide feedback to programme implementers.

The components of the communication strategy[4] were thus (*a*) to segment the audience and identify specific behaviour changes, (*b*) develop message concepts, and (*c*) plan a mix of media based on patterns of media use.

a) Target Audiences and Behaviours: Communication needs assessments were not a routine feature of IEC activities in most states, because the need for such assessment was not recognised, no in-house capacity existed for employing rapid assessment techniques, and no funds were earmarked for this activity. Even when communication needs assessments were conducted in relation to donor-funded projects, the information obtained was not used sufficiently to define segments of the local target audience.

The purpose of audience segmentation was to identify an audience whose behavioural change would affect public health and to determine the best means of reaching such an audience with relevant messages. Target audiences needed to be segmented according to their knowledge, attitudes, beliefs and practices; incidence and severity of the health problem; potential to receive messages as measured by media reach and social networks; geographic location; and likely responsiveness to programme elements.

b) Message Concepts: Message concepts had been aimed at generating awareness rather than at achieving specific goals for changing behaviour. In family planning, for example, messages focused on promoting the norm of small families and paid little attention to method-specific information on the correct use and management of side effects.

The potential role of IEC in increasing contraceptive use was underscored by the results of two recent surveys. The National Family Health Survey (NFHS) (1992–93) found that more than 60 per cent of the reasons for not using contraception among women who could be using family planning were related to perceptions, misinformation and poor understanding of methods. Only 6 per cent of these women were not using contraception because they found it difficult to obtain. Findings from the district-level baseline survey of the FP programme in UP were similar. In Jhansi district, 70 per cent of women were not using family planning because they had concerns about contraceptive methods and their side effects. Only 7 per cent stated that it was hard to obtain FP services.

While general messages do have a role in campaigns to generate awareness, they are less useful for promoting a specific behavioural change. Global experience shows that information about logistics, such as source of supply, correct use and benefits, are directly linked to changes in behaviour.

Materials had to be pretested among the intended audience and not based only on the perceptions of programme staff and managers. Pretesting would ensure that materials and messages were relevant to the specific needs of each target audience segment and that the communication effort had a higher probability of success.

c) Patterns of Media Use and Mix: Effective programmes use a mix of communication channels (interpersonal, group, mass media, traditional theatre) to reach audiences. The goal was to find a mix of channels that could reach large segments of the audience with adequate frequency. The FWP used IPC in three ways: (*a*) through frontline health workers, male and female multipurpose workers, anganwadis and auxiliary nurse/midwives (ANMs) (*b*) through the Mahila Swasthya Sangh (MSS) scheme, and (*c*) through NGOs.

It was recognised that inadequate counselling produced clients who were dissatisfied with the services provided at health facilities and who did not understand key messages or receive adequate information.

Several factors contributed to this situation. First, frontline workers had inadequate training in counselling skills. Second, for a variety of reasons, workers had limited contact with clients and little motivation to spend time on counselling. Third, IEC activities and inputs were not monitored during supervision visits when priority was placed on determining the number of FP acceptors, cases immunised and cases treated rather than the quality of services provided (including counselling) and the level of client satisfaction.

Recognising that mass media have a limited reach for key target audiences in remote rural and poor areas, the government had begun to emphasise the use of IPC channels like frontline workers in the health and related sectors, the use of community-based link workers (public sector as well as NGOs) and rural women's groups like MSS, and the use of folk media including street theatre. The effectiveness of such large-scale approaches was uncertain and IPC channels clearly needed to be improved.

The MSS scheme, which attempted to improve IPC at the village level by developing a network of community leaders who discussed welfare issues with villagers, had not fully succeeded in its mass education and media activities. Married women reported low levels of exposure to the various communication activities that the members of MSS were expected to undertake (Indian Institute of Mass Communications 1993). It was suggested that increasing the frequency of home visits was one way to improve the quality of these efforts.

In India, there was conflicting evidence regarding the reach of mass media campaigns for conveying social messages. However, the private sector in India used mass media extensively to promote its products and services. It had developed a functioning mechanism for planning, buying and monitoring the media, which the government could access to complement its existing in-house capacity for disseminating a wide range of social messages through the Ministry of Information and Broadcasting.

This was the background, available insight and prevalent thinking regarding IEC when the RCH programme was being designed. The importance of IEC was understood and the need to invest in training health workers in communication skills was well appreciated. Involving NGOs and engaging private agencies to create awareness and undertake communication activities to bring about behaviour change was also accepted by MOHFW.

NATIONAL COMMUNICATION STRATEGY DEVELOPMENT FOR THE RCH PROGRAMME

The Government of India launched the nationwide Reproductive and Child Health (RCH) Programme in 1997. At one level, the programme built upon the existing FWP and strengthened the Child Survival and Safe Motherhood (CSSM) component of the earlier programme. The new programme also did away with sterilisation targets to limit fertility and adopted the Target-free Approach (TFA) that had been successfully piloted in Tamil Nadu after the 1994 ICPD in Cairo. Moreover, the adoption of the RCH approach was an affirmation of India's commitment to the ICPD Declaration that called for gender equity and equality in population programmes and asserted the right of every individual, especially women in the reproductive age group, to control their fertility based on informed choice regarding contraceptive measures.

The Reproductive Health (RH) approach that had been hailed as a 'paradigm shift' in population policy could be defined as

> People having the ability to reproduce and regulate their fertility, women being able to go through pregnancy and child birth safely, the outcomes of pregnancies is successful in terms of maternal and infant survival and well-being, and couples are able to have sexual relations free of fear of pregnancy and of contracting disease.[5]

This new approach also meant that the programme was needs-based, client-centred and demand-driven.

The Indian government's RCH Programme also received the support of multilateral donor agencies and during the preparation of the project several studies were conducted to assess the health needs and prevalent behaviours of poor rural women who were to be the main beneficiaries of the programme. Some of the key findings, confirmed by the National Family Health Surveys, were:

- A sizeable proportion of poor women gave birth to more children than they wanted or desired.
- While they knew about contraceptive methods, they did not use the available health services for a variety of reasons.
- There were prevalent misconceptions and fears about side effects owing to inadequate or incomplete information about contraceptive methods.

- There was low awareness and inadequate access to spacing methods of contraception.
- Husbands and other family members often disapproved of family planning.
- Poor quality of care, improper behaviour and non-availability of staff and lack of referral services were some of the critical barriers to the use of public health (MCH) services.

Similarly the surveys also showed that there was pressure on health functionaries to achieve sterilisation/IUD (Intra–Uterine Device) targets; their mobility to cover the service area was limited; there was a shortage of drugs and contraceptive supplies; there was little support from supervisors; and their own clinical IPC skills were inadequate.

In short, there was a need for major policy reform that would address the need to bring about a profound attitudinal and behavioural change among the large number of managers and health care providers such that the RCH Programme would:

- Focus on client needs, particularly on the needs of women, marginalised groups and minorities, and be responsive to them.
- Obtain client feedback and take corrective steps.
- Provide more complete information and better quality of clinical care.
- Move away from a centralised planning and management system to a decentralised and district-based health care delivery system.
- Induct NGOs to support in creating awareness regarding RCH and extend the outreach of services.

One of the key tenets of the RCH approach was Community Needs Assessment (CNA) wherein the health needs of the community were to be assessed in consultation with the community. The ANM would work out the requirements of the community through community meetings, family visits and support from the MSS members, anganwadi workers (AWWs) and panchayat members.

IEC was visualised as an important and integral part of the new RCH Programme and it had to elicit the participation of families and communities in determining the health needs of the community. It also had to make sure that the service provider was equipped with information on the new RCH approach and had been trained with the requisite IPC skills to be more sensitive and responsive to the needs of the more vulnerable sections of society. Above all, the contours of the new RCH Programme had to be communicated and understood by the service providers and client community alike.

In other words, IEC had a small but critically important role in the RCH Programme. It was vital for increasing the demand for services as well as for improving the quality of service delivery. The core objective of the RCH Programme was women's health and client satisfaction. Given the low status accorded to women in India and the indifference of the health care delivery system to their health, achieving the RCH objective would require a 'mindset' change among service providers in particular and society (including women) in general. The credibility of the public health system and the sincerity of the health care workers were in question and the clients (poor, rural women) often hesitated to approach them. This was the backdrop against which the RCH Programme was being initiated in 1997 and IEC had to persuade all those concerned with it to change their attitudes and behaviour and respond to the new RCH paradigm of needs-based service delivery for client satisfaction.

The first task in this regard was to articulate and define a coherent National Communication Strategy for RCH that would address the issue of bringing about change in behaviour among clients as well as service providers. MOHFW started the process of defining the communication strategy for the country's RCH Programme in January 1999. Experts from the field of communications, management, paediatrics, obstetrics and gynaecology, training, health care services and management, donor agency representatives and NGO partners were invited for a national-level workshop to deliberate with government officials and articulate a strategy of communication that would advance the new paradigm for RCH care: 'A new strategy that would make the leap from awareness generation to behaviour change, from being instructive to being empowering, and from taking the generic approach to taking the individualised approach' (MOHFW 2000).

It was acknowledged that achieving the goal of communication (or IEC) for behaviour change would be a difficult and slow process. The continuum of behavioural change portrayed the readiness of audiences to change: from being unaware to being informed, from awareness, concern and knowledge to trying, adopting and sustaining a new behaviour. An RCH communication (IEC) strategy had to be developed that helped people to move from one stage to the next.

A communication strategy that generates awareness is different from one that promotes behaviour change. An RCH approach required identifying priority target audience segments for each component of the programme.

The framework for an RCH National Communication Strategy that was articulated in the national workshop was later discussed in a series of regional workshops and meetings with state government officials, NGO partners and communication and health care management experts. It was through this extended consultative process that the goals of the RCH Communication Strategy and its main tenets were sharpened. It was recognised that the National Communication Strategy could only give a broad thrust in terms of goal, objective and strategy directions and define the roles that the Centre, states and districts would play in implementing the strategy.

The goal of the RCH Programme's Strategic Framework for Communication was to 'encourage individuals, families and communities to make informed decisions concerning reproductive and child health (RCH) through a programme of health communication which facilitates behaviour change'.[6] Significantly, this goal was to be achieved in a systematic and participatory manner through capacity building of all partners and convergence of services among different sectors. The objectives of RCH communication that flowed naturally from the goal statement were:

- To facilitate behaviour change with regard to key RCH indicators.
- To facilitate decentralised CNA, planning, implementation and monitoring of IEC at all levels.
- To provide technical support in building capacity for IEC among service delivery personnel at all levels.

The five key tenets for the implementation of the National Communication Strategy were identified as:

1. Interpersonal communication (IPC) to bring about behaviour change would be the mainstay at the field level, and would encourage greater dialogue on issues of reproductive and child health between individuals within families and communities.
2. Advocacy interventions based on normative research, and including the use of mass media, would be needed to promote societal change with regard to behaviour norms on RCH issues.

3. This would require decentralisation of some responsibilities for IEC to the states and districts from the Centre and consequently the definition of new roles of the three levels.
4. There would be need for increased engagement with the NGOs, Community-based Organisations (CBOs), PRIs and the private sector for social mobilisation and IEC for RCH.
5. There would be a critical need for capacity building to undertake the newly defined tasks and enhance the image of RCH functionaries.

Components of the New RCH Communication Framework

The changing attributes of the RCH Communication Strategy within the country required a change in the responsibilities for IEC action. Hitherto, the directions had come from the Centre and the states (and further down, the districts) just followed the guidelines it laid out. This was convenient and satisfactory for meeting the objectives of IEC for the target-driven FW programme. In the context of the new RCH Programme, with its shift towards client-centred, needs-based, demand-driven health care services, IEC required to be tailor-made for the local context. Hence, the role of the Centre, states and the districts had to be defined afresh.

The district was the natural unit for CNA-based planning for the RCH Programme and the focus for convergence of government and non-governmental efforts to bring about behaviour change through participatory planning and IPC. It was expected that frontline workers, such as the Block Extension Educators (BEEs) and ANMs, would respond to the identified RCH needs of individuals within the community and also work with communities to influence their normative behaviour and encourage the use of community-specific local knowledge and practices to promote behaviour change through the involvement of PRIs and other opinion leaders within the area.

The district would become the central point for the development of an appropriate and flexible action plan to support the communications needs of the community. This would entail the timely distribution of IEC materials; utilisation of local folk media opportunities; coordination with other related departments, NGOs and PRIs; and ensuring that expectations matched the delivery of health care services according to the specific needs of the community. In order to implement such a planned and coordinated set of IEC activities, the development of capacity among the service providers and community volunteers was essential. Skills in IPC, organising and managing local communication events, tracking changes in KAP, and monitoring and evaluating the IEC programmes were some of the areas identified for staff training. It was envisaged that the district IEC team would also participate in the development of state-specific IEC strategy development, share experiences with other districts as well as develop strategies for the urban areas within the district.

States, on the other hand, had to assume responsibility to evolve a state-specific RCH Communication Strategy that addressed the particular information/communication needs and state-specific behaviour norms that emerged from the RCH household survey data. Literacy, educational levels and socio-economic backwardness affected attitudes and practices and communication had to be sensitive to these issues. There were also variations within states, with some districts performing much better and others being under-served for a variety of reasons. Similarly, there were also wide disparities between states and no national strategy could accommodate all the variations. Kerala and Tamil Nadu, for instance, had the infrastructure and facilities to try and achieve the RCH goal of 100 per cent institutional delivery. Rajasthan and Bihar, on the other hand, could perhaps only attempt to reach a target of increased safe delivery by trained birth attendants (TBAs) since the Primary Health Care (PHC) services were woefully inadequate in these states. Thus state communication strategies had to keep in mind the local conditions.

In addition, state-level capacity had to be developed for planning and implementing IEC for behaviour change. This included assessment of current health and family welfare practices; assessment of media channel outreach; pretesting and dissemination of media materials; implementation and monitoring of IEC activities; and evaluation and refinement of IEC programmes. State-level planning was also required to support district-based effort using local folk media and IPC skills. The state-level plans would be informed by the plans of districts (and self-governing urban areas) and the states would monitor the actions taken by the constituent communities and track the effectiveness of such IEC effort on the response to delivery of RCH services.

The states had a critical role in drawing together expertise from government and non-government sectors in the areas of research, training, health care, advocacy and management and develop and undertake concerted state-specific communication campaigns for population and women and child health issues and other components of RCH. They had to become the hub of campaigns that addressed the communication challenges and produced materials that were pre-tested and relevant to the local milieu. They also needed to encourage the sharing of experiences among districts, develop IPC skills among service providers and address the districts' staff and other resource needs. The states were also required to participate in national campaigns and support the development of a National Communication Strategy for RCH.

The Centre was responsible for the overall policy development process for IEC within RCH, albeit with the full involvement of and in coordination with the state administrations. Under the present National Communication Strategy, the Centre would monitor the actions undertaken by states and track the effectiveness of such actions on the changing picture in India in response to delivery of RCH services and communication efforts.

The Centre's main tasks were to provide technical support for state-specific IEC strategies; ensure timely allocation of resources for states/districts; conduct research on communication for behaviour change; develop media materials and establish high professional standards of quality for IEC materials; and provide support for training and capacity building in states and in MOHFW. The Centre also had to undertake communication campaigns to draw attention to such population and maternal and child health issues that required changes in societal behavioural norms. For instance, son preference and early age at marriage were issues that required patient and persistent country-wide effort. New issues like adolescent health, gender-related concerns like sex-determination tests and female foeticide required urgent attention at the national level.

The Centre was required to bring together expertise from the government, voluntary and private sectors in the areas of research, training, health care, advocacy and management and converge the effort of other relevant departments in order to make the National Communication Strategy and action plan more effective. It was suggested that the Centre should support the creation of a clearinghouse for issues related to population and women and child health issues and other components of RCH, which could then serve as a resource for research, training and communication materials development in states and districts within states. More importantly, the Centre was required to enhance the capabilities of relevant staff within the IEC Division of MOHFW (as well as appropriate capacities at state level) to contract agencies for KAP studies, media tracking and evaluation and to support centrally sponsored mass media efforts which addressed overarching issues within population and reproductive and child health.

The RCH Communication Challenge for Behaviour Change

The RCH Programme was a landmark in terms of altering the priority of the FWP to achieve better health and a stable population. As a programme for the delivery of health services, the RCH Programme had three principal components:

- Meeting unmet need for contraception.
- Improving maternal health.
- Raising chances of child survival.

In order to achieve these goals, communication interventions had to bring about behaviour change by addressing the resistances manifest in prevalent social norms and cultural practices. Also, communication had to provide timely and correct information to create a supportive environment for bringing about the necessary behaviour change among individuals, families and communities.

The first National Family Health Survey (NFHS) preceded the launch of the RCH Programme in 1992. A sample RCH Household Survey was also part of the RCH Programme. Consequently, a vast amount of information was available for planning and identifying service delivery needs and communication priorities. During the course of preparing the National Communication Strategy for RCH, a detailed exercise to identify communication priorities was carried out in consultation with experts and based on the survey findings. These helped in defining the broad areas of concern and set down the expected behaviour change indicators for the communication interventions. For instance, field data showed an alarming decrease in immunisation coverage or fall in contraception use in some states. To arrest this fall became an immediate priority for the communication intervention. Data was also available through various KAP and other studies across the country regarding prevalent social and cultural practices that acted as barriers to adoption of new behaviour (like early registration of pregnancy to avoid risks during childbirth and availing antenatal care (ANC) during pregnancy).

Based on the available data and through a consultative process, a generic Matrix of RCH Behaviour Change Objectives[7] was defined that looked at each RCH issue, identified the indicators (or objectives of communication) of expected behaviour change, listed some of the barriers to that change based on available data that became the communication challenge, suggested some possible opportunities 'to trigger change', and prioritised communication in terms of messages and audiences. At the national level, this was the modest goal of the RCH National Communication Strategy and it was hoped that each state would take the 'template' and further refine and adapt the matrix to its needs and priorities that would then get even further sharpened and focused in the district IEC action plans depending on the prevailing conditions.

In developing this matrix, (see Table 7.3) an attempt was made to order priorities in terms of behaviour changes that had to be addressed urgently (within one year) and those that would require more time to become effective (three to four years). For instance, arresting the drop in immunisation required urgent action through a concerted effort at publicising the immunisation schedule and intensifying coverage through house visits and persuasion using IPC. Certain other societal normative issues like preference for sons and early marriage of girls had to be addressed at a deeper level of social and cultural values through an engagement with families and communities and it was expected that the change would take place slowly over a long period of time. However, while the results of the communication endeavour in the three cases might show up at different points in time, the communication effort for all three had to be initiated simultaneously. Planners and managers were required to allocate budgets after ascertaining the relative cost-effectiveness of the possible interventions and assign responsibility for designing messages and dissemination accordingly.

Although the National RCH Communication Strategy Framework outlined distinct roles for the Centre, states and the districts, issues relating to behaviour change had to be addressed at all three levels. The Centre was better placed to sustain a long-term campaign to address societal issues or advocacy interventions through the mass media in a cost-effective manner. The states, on the other hand, could address regional specificities and plan their campaigns accordingly. Districts could match communication with service delivery and reach out to individuals, families and the community through IPC and traditional folk and other local media channels and group interactive processes. Efforts at all three levels were meant to support one another and ensure that there was no dissonance in the messages that went out from different sources.

IMPLEMENTATION OF THE RCH NATIONAL COMMUNICATION STRATEGY[8]

The objective of the IEC component in the RCH Programme was to improve health and care-seeking behaviour among target populations and thereby increase demand for services and community participation and responsibility for RCH. The thrust of IEC was to improve the IPC skills of frontline workers in order to motivate clients and organise social mobilisation at the field level to ensure community support for RCH. It was felt that there was a need to focus on behaviour change, identify the various audience segments, refine the communication approaches to each segment, and integrate better monitoring and evaluation of IEC *and* of behaviour change.

At the national level, the main focus was to create an enabling environment for RCH with its 'paradigm shift' to create a more client-centred, enhanced quality of care and target-free approach. This was to be achieved by strengthening the quality of mass media approaches to build a supportive social climate for behaviour change. At the state and district levels, the focus was on identifying and responding to local needs. Each state was to develop an IEC strategy and each district would eventually be required to develop an IEC plan of action and implement and assess the effectiveness of the plan. At the frontline, the focus was on better quality IPC and counselling skills and substantial training was imparted in this direction by the project.

Table 7.3
Matrix of RCH Behaviour Change Objectives

RCH issue	Indicators (objectives) of expected behaviour change	Barriers to behaviour change: communications challenges	Opportunities to trigger change	Communications priority
Unwanted fertility	Arrest fall in contraception use in some states***	Social and religious disapproval Non-availability and poor quality of contraceptives	Availability of easier, safer methods	Inform and motivate regarding availability of both spacing and terminal methods, especially in those states with fall in use***
(a) *Unmet need for contraception*	Effective contraceptive coverage by 75% couples**	Perceived costs/risks of contraception	Non-scalpel vasectomy (NSV) for men	Reach out to men about NSV and male responsibility**
	Increased use of contraception among couples with less than two children**	Husband's lack of support and/or fertility preference	Improved husband–wife communication	Focus on spacing between first and second child
(b) *Unsafe abortion*	Drop in illegal abortions*	Social stigma, fear and lack of counselling/guidance		Widely publicise legal provision and hazards of illegal abortion** Promote counselling and reach out to adolescents**
	Address son preference* End illegal abortions*	Ambivalence towards further childbearing		Promote gender equality* Promote responsible parenthood and small family*

Table 7.3 (Continued)

RCH issue	Indicators (objectives) of expected behaviour change	Barriers to behaviour change: communications challenges	Opportunities to trigger change	Communications priority
Maternal health	Reduce maternal mortality/ morbidity**	No perception of pregnancy as a time for special care	Motivate service providers through RCH training	Communicate ANC essential schedule***
		Skewed gender relationships	Reach out and involve private practitioners	Communicate risks and dangers of pregnancy and importance of early registration***
(a) High maternal mortality	At least 3 ANC visits during pregnancy**	Inadequate knowledge of risks during pregnancy	Address RH issues with women's self-help groups, MSS, etc.	Inform about availability of ANC and other services and timings to pregnant women, husbands and other members of the family***
	Reduce delivery by untrained persons**	Care-seeking behaviour not internalised		
(b) High maternal morbidity	Universalise registration of pregnancies in all districts*	Lack of information regarding available facilities and service delivery.	Improved husband–wife communication	Motivate service providers, women's groups and other community groups to early registration of pregnancy**
	Improve registration of pregnancies in remote and tribal areas*	Inadequate service delivery staff and lack of motivation	Involvement of community in primary health care through panchayats	
	Strengthen male involvement*	Lack of knowledge and concern	Address RH issues and family responsibility	Build on men's approval of family planning**

(Continued)

Table 7.3 (Continued)

RCH issue	Indicators (objectives) of expected behaviour change	Barriers to behaviour change: communications challenges	Opportunities to trigger change	Communications priority
	Improve maternal nutrition** Reduce nutritional anaemia among pregnant women**	Negative cultural norms and dietary practices during pregnancy	to men's gatherings/ groups	Reach out to young unmarried men** Counter myths and negative cultural and dietary practices**
(c) Adolescent reproductive health	Increase knowledge and awareness on RH issues among adolescents** Delay age at marriage*	Cultural norms	Non-formal education and health camps for adolescents	Improve knowledge and awareness of RH issues among adolescents** Engage community on issues of early marriage, women's health and status in the family*
(d) RTI/STI care	Make RTI/STI a public health priority among medical practitioners** Increase in numbers of cases reported and treated*	Lack of knowledge and perception of illness among women Indifference and misapprehension among medical practitioners	Peer education through women's groups Address issues through medical associations, family health camps	Bridge the knowledge gap among private practitioners*** Encourage husband–wife communication on RH issues**

POPULATION: BRINGING ABOUT BEHAVIOUR CHANGE

Table 7.3
(Continued)

RCH issue	Indicators (objectives) of expected behaviour change	Barriers to behaviour change: communications challenges	Opportunities to trigger change	Communications priority
Child health	Reduce/eliminate neonatal mortality (focus on girl child)*** Eliminate incidence of polio***	Lack of information regarding risk and failure to recognise danger signs and promptly seek care Lack of information regarding availability of services Lack of motivation	Care givers respond to issues that relate to the health of their children	Inform and educate mothers on risks, danger signs and availability of services*** Persuade family and community to end unequal treatment of girls and boys**
(a) *IMR rising in some districts*	Reverse drop in IMR***	Lack of information and motivation De-motivated staff		Motivate service providers*** Motivate family and community on child survival risks and availability of essential services***
(b) *Drop in immunisation coverages*	Arrest drop in immunisation coverage*** Universalise registration of births** Increase percentage coverage of measles immunisation**	Failure to complete the schedule within the timeframe Lack of information regarding necessity of registration		Widely publicise information on complete immunisation schedule*** Motivate service provider and community to register births in remote (poor) areas*

(Continued)

Table 7.3
(Continued)

RCH issue	Indicators (objectives) of expected behaviour change	Barriers to behaviour change: communications challenges	Opportunities to trigger change	Communications priority
(c) Childcare, especially newborn care	Increased number of children with severe ARI being treated and reported** Diarrhoea treatment and prevention***	ARI not seen as risk to child survival Failure to hydrate sick child adequately Girls get less attention than boys		Improve knowledge base of service provider and community regarding risks of ARI*** Communicate home-based treatment of diarrhoea*** Address issue of equal treatment to girls and boys with couples and families**

Source: MOHFW 2000.
Notes: ***Immediate (within one year) term
**Medium term (3–4 years)
*Long term

While the first task for MOHFW was to develop the IEC strategy through consultations with experts and representatives from the states, some implementation of IEC activities was initiated by MOHFW in 1998 with regard to mass media and social mobilisation. These were the development of an interactive panel discussion TV programme, a set of TV spots on the new focus of RCH issues, and hoardings. Also, a major initiative was taken to involve the Zila Saksharta Samitis (ZSSs) or District Literacy Committees to use local media and community interactions on RCH issues in order to mobilise community participation and support for the RCH Programme. MOHFW also invited eminent feature filmmakers to produce feature films with a RCH perspective. These films were to be screened for general audiences.

There were several constraints that existed in the health care delivery system that prevented effective implementation of the RCH

Programme. First of all, communication had to ensure that policy-makers, administrators and programme managers were familiar with the paradigm shift in the programme and were tuned in to the new client-friendly, needs-based approach. Unfortunately, barring the target-free approach and inclusion of a few new components like RTI/STI treatment and adolescent health in the existing programme, many health professionals were unaware of the changes in

Figure 7.1
Poster Promoting NSV for Men

Figure 7.2
Attempt to Evolve the 'Red Triangle'

the RCH approach. It was imperative, therefore, that all health professionals be given a comprehensive understanding of the shift in the RCH Programme and an orientation programme initiated for policy-makers as well.

Since the new approach depended on the service provider's attitude towards the client, it was equally important to ensure that IEC was integrated into the training of all service providers. This included training in IPC skills for both frontline workers and supervisory staff at the PHC level, and planning and executing IEC activities in the field as well as tracking and monitoring them. All service delivery personnel had to recognise that bringing about behaviour change was not the responsibility only of those designated for IEC but a concern of all health care service providers. The general impression that IEC was about media materials (posters and charts) or organising field publicity events or press briefings had to be countered with an understanding about audience segmentation, identifying resistance to adoption of new behaviours, designing a communication strategy, developing key messages, using appropriate media to reach the audience and monitoring the effectiveness of IEC. Training in IEC was also important because IEC had to be linked with the availability of health services in the field. Using mass media and sending messages that were not supported by availability of services in the villages went against the grain of the programme. Hence, the emphasis had to be on local conditions and IEC had to be planned and implemented bearing in mind local cultural practices, health priorities and service availability.

The planning of IEC, however, had always been centralised, with MOHFW setting down the budget allocations for different activities. States had limited flexibility and the role of districts was reduced to only receiving materials. Field publicity programmes were organised at the state level without much consultation with the district health staff. Often this lack of coordination resulted in wasteful expenditure as print materials remained unused and shows and interactive group meetings did not enthuse the frontline workers or relate to the field reality regarding availability of services. Moving to a more decentralised system where the planning and implementation was done at the district level meant an alternative budgetary allocation with greater flexibility for the states and districts to plan their IEC activities and simultaneously building capacity at the state and district levels to implement such a decentralised IEC action plan. In the course of the RCH Programme several attempts were made to implement the new IEC strategy with its focus on demand generation that was needs-based and client-friendly, and had the community's support. Some of these experiences are examined here with a view to understand the difficulties faced by the RCH Programme.

Social Mobilisation: The ZSS Initiative

Soon after the launch of the RCH Programme, it was decided by the MOHFW to induct the Zila Saksharta Samitis (ZSSs) to create awareness about RCH and generally take charge of field-level social mobilisation for RCH. As the ZSS had done good work in social mobilisation for the Total Literacy Campaigns (TLCs) launched by the Ministry of Human Resource Development (MHRD) in several districts, it was felt that the same structure would be useful for creating community support for RCH. The Secretary, Department of Family Welfare, Government of India, wrote to all the District Collectors, who were also the Chairpersons of the ZSS, inviting them to send in proposals for social mobilisation to create awareness about the new RCH Programme.

The expectations from the ZSS, as per the letter from MOHFW, was that the Collector as the Chairperson of the ZSS, would seek the assistance of the District Health Officers (DHOs) or the Chief Medical Officer (CMOs) to prepare a calendar of events for creating awareness regarding the new RCH approach and mobilise the community towards meeting the objectives of the programme. The Health and Education Departments in the districts were expected to prepare the proposals together (based on local needs and conditions) and the synergy was expected to benefit the RCH Programme through inter-departmental collaboration and effective community participation.

Funds would be provided directly to the ZSS through the State Family Welfare Departments and a provisional allocation of Rs 500,000 per district per year for two years was made by MOHFW. A joint committee of MOHFW and MHRD was set up to scrutinise and sanction the proposals received from the ZSS.

Initially, not many proposals were forthcoming and those that were received were very sketchy. The scheme therefore took quite some time to get off the ground. The ZSS Secretary did not have access to the Health Department officials and the District Collector did not give it enough priority. The DHO/CMO were not always cooperative and the State Health and Family Welfare Departments were indifferent to the scheme since they had not been consulted prior to its launch. Orientation meetings with the State Resource Centres (SRCs) under the National Literacy Mission (NLM) were arranged to help the ZSS submit proposals. The whole process was flawed and dogged with delays at every stage. Weak proposals were submitted after much persuasion. There was delay in sanctioning the proposals and further delay by MOHFW in releasing the funds. The funds were then held up in the State Department of Family Welfare and, in some cases, were sent to the DHO/CMO instead of the ZSS. Nevertheless, by November 2000, 227 districts (ZSS) had been given funds ranging between Rs 300,000 and Rs 500,000 per district and totalling Rs 89.6 million. After that, MOHFW stopped the scheme. However, it did initiate an evaluation of the ZSS experience through the Population Research Centres (PRCs) in six states where the scheme had been operational.

The evaluations[9] showed that in several districts there was hardly any IEC activity because the ZSS itself was inactive and did not have any outreach in the villages. By the time the RCH scheme for ZSS was launched in 1998–99, the Total Literacy Campaign was over in many districts and whatever wave of enthusiasm it had generated had died out. In some districts there was no literacy activity and in others the post-literacy phase had started in classes with neoliterates. There was no 'mass contact' with the community as had been in the TLC phase. In some districts, the Secretary of the ZSS received the funds and handed them over to the DHO/CMO who did the IEC work. In these districts, the proposals had been drawn up by the DHO/CMO in any case since the ZSS was virtually inactive. The long delay in receiving funds eroded whatever interest was there in the scheme and IEC happened in a routine manner. There was none of the campaign spirit that was evident during the TLC phase. The evaluation showed that only in Himachal Pradesh did the two departments work together, with the ZSS being equally active in organising exhibitions and folk drama performances in the villages to generate community support for RCH. This was probably because the NGO partners of ZSS continued to be active in the state.

The community members hardly knew about the special effort for creating awareness regarding RCH that was being undertaken by the ZSS. Most of them said that they had got their information from the frontline workers and, to some extent, from TV spots. Wall writings, posters and folk drama performances and group meetings organised by the ZSS were too few and far between to have any significant impact. In any event, the issues taken up by ZSS were the usual family welfare issues of the old programme as the understanding on the RCH approach was absent.

In effect, as a supplement to the DFW's own effort, the ZSS venture remained a weak attempt to create awareness about RCH. It was not an intensive communication campaign to mobilise the community, as had been the case with TLC. Also, during the TLCs, the environment building or social mobilisation effort was immediately followed by enlisting volunteer teachers and identifying adult learners so that literacy classes could commence. In the RCH Programme there was no such direct linkage and the diluted effort only made a marginally incremental contribution, if any at all. Since the initiative had been taken by MOHFW without consulting the state governments, there was some resentment and general indifference towards the scheme. The logic of inducting a sister department to achieve a common goal was not appreciated in the field as several health functionaries felt that they were being bypassed and their competence was being questioned. The weak presence of ZSS in most districts only made matters worse and showed up MOHFW's faulty planning.

Since the ZSS scheme was a failure, some means of launching a social mobilisation campaign in the districts to generate demand for quality RCH services remained an important concern. It was decided that the planning for this should be left to the districts, allowing them the flexibility to design the campaign according to the resources available locally with some outside support.

IEC for RCH Scheme in States

The National Communication Strategy for RCH emphasised the need for decentralising some responsibilities for IEC to the states. This need was also borne out by the poor performance of the ZSS scheme owing to lack of supervision and ownership by the states. In view of this, a scheme was envisaged as part of the plan to restructure the RCH Programme during the mid-term review held in November 2000. Accordingly, funds were allocated to low-performing states to execute IEC activities over the remaining period of the RCH project.[10]

The scheme presumed that the communication activities would be based on a specific strategy and the states would develop their action plan following the priorities highlighted in the National Communication Strategy. A workshop was held with representatives from the weak-performing states and it was made clear that the states had the flexibility to use the funds in any manner they chose. However, they were expected to define a communication strategy based on the RCH issues and priorities of the concerned state. The RCH Household Survey data was to be used to define these priorities and since funds were limited, the states were advised to take up one or two issues and spread the effort in a few districts so that the intensity of the communication campaign would not get diluted. Financial support was provided for a range of IEC activities based on the priorities defined by the states. These activities can be grouped under the following broad heads: *a*) mass media—radio and TV dissemination (with limited production activities) through AIR and Doordarshan; *b*) outdoor publicity, mainly hoardings, bus panels and other urban outdoor media vehicles; *c*) rural publicity through wall writings, posters, exhibitions, banners and displays in fairs and festivals, and folk media performances; and *d*) interactive group processes like orientation camps, community events and group meetings. The only binding condition was that the funds used for mass media should be limited to 30 per cent of the budget allocation. A small, initial sum amounting to Rs 4.5 million for one year per state was released in March 2001 as an experiment to assess the states' capacity to plan and execute an IEC campaign. Subsequently, this approach was taken further under the special scheme for the Empowered Action Group (EAG) of states and Rs 30 million for smaller EAG states and Rs 50 million for larger EAG states were allotted in 2002–03.

The experience was not entirely satisfactory as the states lacked adequate understanding and/or interest as well as the professional support to plan and execute the scheme. Orissa was the sole exception (Box 7.1). MOHFW was expected to provide the states with technical support to develop the IEC strategy and a campaign/plan for utilising the funds. For a variety of reasons, MOHFW did not provide the necessary support. The states themselves were not able to do much. Under the scheme, Hindustan Latex Limited (HLL) and Federation of Indian Chambers of Commerce and Industry (FICCI) were engaged to help states develop the IEC action plan. Progress with that was tardy and after a while FICCI withdrew from it altogether.

Box 7.1
Orissa—An Exception[11]

Orissa had taken a conscious decision to restrict the implementation of the RCH Special Scheme for IEC to five districts and focus the campaign on Antenatal Registration and Care (ANC) since the Maternal Mortality Rate (MMR) was fairly high in the selected districts. While a combination of mass media (radio and TV), outdoor and print publicity (hoardings, wall writings and posters), video shows and street theatre performances were used during the campaign, capacity building of frontline functionaries like male and female health workers, anganwadi workers, traditional birth attendants with IPC and counselling skills and focusing on the importance of ANC and spacing of births after the first child was undertaken during the campaign. A key message to frontline workers was the importance of building relationships with the expectant mothers and their families and that became the basis of further IPC to ensure regular ANC check-ups and availing of services.

Portable exhibition kits had been prepared on the theme of ANC and these were supplied to all the block PHCs and were used extensively during the training programmes. Mahila Swasthya Sangh (MSS) meetings were organised regularly, with video films being screened during these meetings for mothers. One of the films produced under the RCH Special Scheme for IEC was on the theme of sex-determination of the foetus and the film countered the belief that the mother was responsible for the birth of a girl-child.

(Continued)

> **Box 7.1**
> *(Continued)*
>
> The other important thing was that tracking and monitoring was an integral component of the IEC implementation plan. A regular time schedule had been drawn up with the District MEIO (Mass Education and Information Officer) being responsible for the IEC campaign in the district and the BEE being responsible for the effort in the blocks. A unique feature of Orissa was that none of the sanctioned posts for BEEs were vacant. This gave strength to the IEC campaign, as there were persons who could responsibly execute the implementation plan. What was most interesting (and very different from other states) was that the BEEs were relatively young, motivated and enthusiastic about IEC. All of them had been trained at the National Institute of Applied Human Research and Development (NIAHRD), Cuttack, under the Specialised Training in IEC under RCH implemented through the National Institute of Health and Family Welfare (NIHFW). The training had been quite thorough and successful and the BEEs found the field visits to equip them with research tools for conducting CNA, pretesting media materials and assessing effectiveness of a media campaign exhibited by them during the training very useful.
>
> During a field visit to the Bentakar block of Cuttack district, it was heartening to note that the importance of IEC to bring about behaviour change was appreciated and it was in a spirit of teamwork and cooperation that a field-level IEC strategy was worked out. For instance, since ANC registration was the focus, concerted effort was made to use all channels available for the different target groups. Regular monthly meetings were held at sector PHCs and then at the block PHCs (supervising the three sector PHCs). Once a month all BEEs met at district headquarters and every month, all District MEIOs met at the state headquarters in the State Institute of Health and Family Welfare (SIHFW). This helped in identifying problem areas, whether relating to service delivery, medical supplies or determining the focus and stress of IEC. The close cooperative relationship was evident and the ANC registration had increased after the IEC campaign through the different media, including MSS meetings, capacity building of frontline workers, local media performances and orientation programmes for Gram Panchayat members. The Medical Officer (MO) at the PHC appreciated the IEC effort and clearly understood its importance in RCH. The BEEs were aware of the statistical information/data for the block and knew that the IEC effort had to be fine-tuned

> **Box 7.1**
> **(Continued)**
>
> to the needs and requirements of the particular area and seasonal occurrence of diseases.
>
> The 'links in the chain' for the IEC planning and implementation process from the sub-centre, sector and block PHCs to the District FW Bureau and the state headquarters were maintained through regular monthly monitoring meetings on IEC. The close cooperation and understanding between the medical personnel and IEC staff at all levels was also evident. The importance of capacity building of frontline workers in IPC skills, emphasis on area-specific planning for IEC based on local health status and problems, the functioning of community-based groups like the MSS (which are mainly women's self-help groups [SHGs] in Orissa) and a coordinated IEC effort along with tracking and monitoring appeared to be well understood. ANC was clearly the focus for IEC at that time and the latest survey data showed increased ANC registration as well as higher numbers of institutional deliveries. The BEEs naturally felt pleased and enthused by the success of their effort.

In October 2000 when the National Communication Strategy was approved,[12] it was agreed that a pilot project in 10 districts across the country (where externally aided programmes were ongoing) would be tried whereby a social mobilisation campaigns would be launched in each of these districts with the collaboration of other development departments and NGOs, etc. The European Commission (EC) agreed to provide the additional funding support that was required for this effort. This was planned as a one-year pilot project and districts were identified for the purpose. Unfortunately, despite effort by the IEC Division, MOHFW and the EC, including visits to some states and holding meetings there, not even one pilot project was initiated. This was very disappointing and pointed to the fact that the states, particularly the low-performing EAG states, had no understanding of the importance of IEC (termed Behaviour Change Communication [BCC] in RCH II). Clearly, the states' capacity for IEC (BCC) planning and strategy development needs to be improved in great measure and MOHFW will have to take this up urgently and provide the necessary support for it.

Capacity Building

Training or capacity building of all categories of health service delivery staff was recognised as essential in the design of the RCH Programme. The new approach, with its emphasis on service quality based on client needs and demands, required a change in attitude. The focus on women's health and nutrition required improvement in the clinical skills of frontline staff as well. To make the delivery system more efficient and responsive to client needs meant better management and induction of new management skills. In order to bring about an attitudinal change among the service providers, the National Institute of Health and Family Welfare (NIHFW) launched a nationwide orientation and training programme for RCH, of which communication or IEC was an integral part. However, this component was different for different levels of staff: for instance, for frontline workers the focus was more on IPC and initiating group processes (for conducting CNA), while planning and managing local IEC initiatives formed the thrust for the PHC and district-level staff. The magnitude of the training effort was overwhelming—280,000 service providers needed to be trained while the programme was already on the ground. The process was not only time-consuming and somewhat ineffective because developing the curricula and training materials, identifying training institutions, building their capacity, and persuading state governments to depute staff for the training were all difficult and tedious tasks. NIHFW was hardly equipped to fulfil its role as coordinator and supervisor in a satisfactory manner. The imperative was to somehow complete the various training programmes, which not only took an inordinately long time, but also neglected some components in the process. Among these, the IEC or communication component was the most affected. Service providers regarded training in clinical skills and RCH programme management as more important and necessary in their work.

Capacity building of IEC personnel at all levels was another significant initiative in RCH. Two training programmes (January and July 2000) in IEC planning and management were conducted at the National Institute of Design (NID), Ahmedabad, for MOHFW and state-level IEC officers.[13] The curriculum for this training programme was developed in consultation with MOHFW and focused on the planning and management of a strategy for Behaviour Change

Communication, which was the new thrust in RCH. Contracting outside agencies, case studies of successful, field-level communication efforts, importance of communication research to identify the felt needs of audiences pretesting media materials, and tracking and monitoring IEC initiatives were some of the components in the NID training programme that was conducted in a participatory framework.

NIHFW also developed curricula for district and block-level IEC personnel for specialised training in IEC. For the district IEC officers the training was for one week while BEEs underwent a two-week residential programme. NIHFW developed the curricula for these trainings in consultation with experts and with the participation of the identified training institutions. Both the training programmes were designed as a mixture of fresh inputs on the communication challenges in RCH and practical guidance in planning and executing a communication programme using locally available media. The training components included audience segmentation, message design, scheduling of activities, budgeting, tracking and monitoring. While the emphasis for the district-level staff was on management of local media including print and folk media, the BEEs were trained to support the frontline workers in conducting CNA through household visits and community meetings. In both training programmes the emphasis was on practical aspects—visits to the community and understanding community needs, developing simple messages and media materials and assessing their effectiveness in the community.

Unfortunately, NIHFW took a long time to finalise the curricula and training materials. Identifying appropriate institutions with facilities and faculty to conduct the specialised IEC training was also not easy. It was only by the end of 2000 that the first trainings were held for district IEC officers. Though the training design was fairly comprehensive, it was dependent on the capability of the faculty in the identified training institutions. Many of the institutions identified by NIHFW did not have adequate experience in IEC and NIHFW was not able to provide technical or professional support to them.

In addition, most states had little interest in IEC and many posts for IEC staff were vacant. The training institutions often found that the number of participants attending the training programme was far lower than had been estimated. Even after persuasion by NIHFW[14] many training courses remained poorly attended. By the end of 2003 nearly 50 per cent of IEC personnel at both district and block level, remained untrained. This was a sad comment on the status of capacity building for IEC in RCH and showed the overall lack of interest in IEC in the states.

The Centre for Media Studies (CMS) conducted an evaluation of the IEC training programme.[15] In its report, CMS pointed out that the institutions identified for executing the training were not adequately equipped to conduct specialised IEC training. Some of them had some experience in health and family welfare but not in IEC. Even specialised institutions like the Indian Institute of Mass Communication (IIMC) did not conduct the training satisfactorily. While the curricula was well thought out and flexible, and could be easily adapted to local specificities, the training institutions (with the exception of CHETNA, Ahmedabad) treated the training in a perfunctory manner. NIHFW did not provide enough supervision. The worst affected appeared to be the low-performing EAG states where the need for training was more acute. CMS also questioned the wisdom of identifying training institutions from outside the state. Had institutions from within the state been chosen, the training could have been more locally specific.

Engaging Private Sector Professional Agencies

Consistent with the effort of developing the RCH Communication Strategy, early initiatives were taken by MOHFW to induct private sector professionals to produce materials for the mass media. Feature films, radio serials, interactive panel discussions, TV spots and hoardings were some of the activities that were tried under RCH. However, these initiatives were not planned as part of any comprehensive communication strategy. The series of interactive panel discussions for television with a popular actress hosting the show did not hold audience interest. The producer had not done adequate research and relied on the usual format of an expert answering questions from an invited audience. The popularity of the TV host was not sufficient to sustain the interest of viewers. Similarly, MOHFW commissioned eminent film producers to make feature films based on RCH issues. Some of these films were of good quality and even won national awards.[16] The large investment required for producing feature films was justified because of the higher production value for films as against the earlier experience of producing a large number of average quality telefilms. No thought was given, however, to the need for commercially distributing or exhibiting the films. The National Film Development Corporation (NFDC) made some attempt to do this but with limited success. Thus, hardly any of the mass audiences that go to see commercial feature films got a chance to view these films. Consequently, the key RCH issues of women's health and nutrition, violence against women and gender equality that were covered in these films did not get sufficient attention or exposure and the health service providers continued to neglect these areas of concern.

Numerous TV spots were produced but these were made in an ad hoc manner on a series of RCH topics without specific targeting of specific audience groups. The briefing by MOHFW was inadequate and there was no control over production quality and time schedules. The lesson is clear: just engaging private agencies cannot yield satisfactory results unless the hiring agency is well prepared and clear about what it wants and has the ability to ensure that the hired agency delivers a satisfactory product.

Towards the end of the first phase of the RCH Programme (RCH I), MOHFW engaged an advertising agency to do its mass media buying for TV. Since the expenditure on this activity was to be reimbursed by The World Bank, elaborate procedures were adopted to ensure transparency in the selection of the agency. All this took a very long time of over a year. The agency's job was limited to make a 'best value for money' selection of TV programmes so that maximum exposure for TV spots could be ensured for the target audience. Since the agency had no control over the quality of the individual spots and had to choose from those that were available with MOHFW, the overall result was not very satisfactory. Apart from transparency in the selection procedure (an achievement in itself), the only other benefit was the periodic monitoring reports provided by the agency that gave some indication of the awareness levels and understanding of RCH issues through audience recall of the TV spots. But since the spots were not linked to any service delivery initiative (like a special drive or campaign), there was no way of ascertaining their effect on behaviour.

The unfortunate implication of this experience with private agencies was that government officials began to think that the departmental agencies under the Ministry of Information and Broadcasting were less expensive and more tuned in to government requirements. The realisation that the failure to utilise the private agencies lay in the poor briefing and inadequate preparedness to deal with them was not understood or admitted. There was no appreciation of the importance of research and proper planning or the ability to use the creative potential of private agencies to best advantage. It is always more worthwhile to engage an agency to do the research, planning and development of a strategy, prepare creative media materials and recommend a media plan and also provide monitoring reports assessing the effectiveness of the exposure. By doing this, the agency becomes responsible for introducing professionalism in every stage of the planning, design and execution of a communication initiative. The health professionals and MOHFW's IEC Division are then only left with the responsibility of providing a proper brief, ensuring quality of design and execution, alerting the field functionaries and assessing the impact. The last aspect is very important since mass media only give a very broad cover. The states and the districts have to be geared to use their frontline workers to take the mass media messages forward with field-based IEC activities using local folk media, interactive group processes and IPC. This is essential because then IEC becomes localised, relevant, appropriate and in consonance with local conditions and the services on offer.

EMERGING ISSUES FOR COMMUNICATION PLANNING

The Government of India and the external donor agencies supporting the RCH Programme carried out an assessment of its performance since it became operational in October 1997. Based on this, they have attempted to change and improve the design of the second phase of the programme (RCH II) that is now operational. In the meanwhile GOI announced its National Population Policy (NPP) in 2000 and its National Health Policy in 2002. The latest Census of India in 2001 has also thrown up significant information on the demographic trends and indicators regarding the stabilisation of India's population.

The NPP (2000) affirmed India's commitment to the International Conference on Population and Development (ICPD) Programme of Action (POA) and asserted the centrality of human development, gender equity and equality and adolescent reproductive health, among other issues, to stabilising India's population. This move away from the earlier target-oriented, coercive programme of population control confirmed GOI's faith in the new RCH Programme that focused on the special concerns for women and child health and improving both the quality of care and access to health care facilities. The NPP has been emphatic in stating that

> ... stabilising population is not merely a question of making reproductive health accessible and affordable to all but also increasing the coverage and outreach of primary and secondary education, extending basic amenities like sanitation, safe drinking water, better nutrition and housing and empowering women with enhanced access to education and employment (National Population Policy 2000).

The NPP (and the National Health Policy 2002) laid emphasis on decentralisation of planning and implementation that promotes demand-driven, area-specific integrated and high quality reproductive and child health care services.

On the other hand, the Census of India 2001 showed that India's population continued to grow and had crossed the one billion mark. Though there were vast regional differences with some states having achieved the net reproductive rate (NRR) of one, the main underlying reasons for continuing population growth rate was attributable to three factors:

- Large size of the population in the reproductive age group (estimated contribution 58 per cent).
- Higher fertility due to unmet need for contraception (estimated contribution 20 per cent).
- High wanted fertility due to the high IMR or infant mortality rate (estimated contribution 20 per cent).

Added to these was the fact that over 50 per cent of girls got married before the age of 18, the legal age of marriage, resulting in a typical reproductive pattern of 'too early, too frequent, too many'.[17]

The Census of India also showed that the sex ratio in several states, particularly in the age group 0–6 years, has been worsening for girls, indicating that 'son preference' and regarding the birth of a girl as a burden to the family are a continuing cause for grave concern. The easy availability of diagnostic tests to determine the sex of the foetus, even though illegal under the PNDT Act 1994, and subsequent female foeticide is common in many parts of India even though it is a criminal offence. Gender injustice begins at the foetal stage and continues through discriminatory childcare practices leading to impaired growth and development of the girl child. Educational inequalities between boys and girls are pronounced, with girls having a much lesser chance of being admitted to school and/or completing their basic education. Adolescent marriages, illegal under the Child Marriage Restraint Act of 1978, continue to be prevalent in much of rural India (even though the mean age of marriage for girls has been rising and stands at 20 years for all of India). Adolescent marriage is synonymous with adolescent fertility. Teenage pregnancies are not safe and adolescent mothers are twice as likely to die from complications during pregnancy as compared to women over 20 years of age.

The MMR continues to be shamefully high in India and four mothers die per 1,000 live births each year in India. It is estimated that for every woman who dies during childbirth, as many as 30 other women develop chronic, debilitating conditions that seriously affect their quality of life. Skilled personnel attend to only 35 per cent of deliveries in India and in some districts only 5–10 per cent deliveries are conducted by skilled personnel. Nearly seven million abortions take place annually in the country, and for each legal abortion, it is estimated that another 10 illegal abortions take place but go unrecorded. Nearly 12 per cent of all maternal deaths are attributable to abortion-related complications. Maternal mortality is high in the states where fertility is high, simply because women there are having more children. It is also high in those states where children are born to very young women and to women who have multiple, closely spaced pregnancies.

The National Family Health Survey II (1998) and other surveys have indicated that unmet demand, both for limiting and spacing, continues to remain high in many states. In UP, 38 per cent of couples have expressed unmet demand for limiting family size (birth) and 18 per cent want to postpone the birth of their next child for another two years, but are still not using any contraceptive methods for want of proper information and services. Only 16 per cent of women below 30 years of age practice contraception for spacing. Male participation in accepting responsibility for contraception is woefully negligible— only 1.9 per cent of total sterilisation acceptors adopted vasectomy in 1997—and the entire burden of contraception falls on the women.

Women's empowerment through which women can achieve greater equality with men is consequently still a distant dream in India. The government, despite being a signatory to many international covenants and policy pronouncements (including the NPP and the RCH Programme), has been unable to extend fundamental social, economic and political rights to women. At a societal level, women continue to suffer oppression and injustice that sometimes take a vicious and violent form, at home, in the community and in society. Obviously, the paradigm shift enunciated in the RCH Programme has not been understood or assimilated even among the service providers, leave alone society at large.

The task of advocating for women's equality and equity, reproductive rights, enhanced quality of care, adolescent health and nutrition and client satisfaction and bringing about change in behaviour among the service providers and clients alike, remains an unfinished agenda of the RCH Programme. Even though there has been a fairly consistent articulation of the National Communication Strategy for the RCH Programme, its implementation has been poor. If the programme is to become more effective in its second phase (RCH II), some of the reasons for the failure of communication or IEC/BCC will require to be redressed.

One major problem of IEC seems to be the lack of professional and trained staff at the state, district and block levels. Without adequate staff and lack of promotional avenues, most IEC personnel at the state and district levels have become discouraged and lack the motivation and direction for improving their performance. Without leadership and direction set by a professional team at the headquarters, IEC becomes a routine activity of production and dissemination of media materials. The linkage between the health professionals responsible for delivery of services and IEC personnel is often weak and appreciation of IEC is lacking among the health professionals.

Even at the state level, the position of the Joint/Deputy Director (IEC) is often held by a doctor and not by a professional IEC person. The Rajasthan IEC Bureau model or the Orissa model of IEC being located in the State Institute of Health and Family Welfare (SIHFW)—away from the DFW and linked to the training institution—has certain advantages that should be examined for other states. Only in Orissa have all the sanctioned BEE posts been filled and the newly appointed BEEs have benefited from IEC training. Most other states have a large number of vacancies and many BEE posts have been abolished. Staffing is a problem that requires urgent attention, especially at the district level and below.

The existing IEC staff was recruited when the IEC Division was still being called the Mass Education and Media (MEM) Division. A range of technical personnel responsible for production and distribution/dissemination of media materials—photographers, artists, projectionists, etc.—make up a fair number of IEC personnel at the Centre, state and even district level. Most of them have a limited technical function and retraining them (as has been attempted in RCH) is possible only up to a certain extent. With the new thrust towards behaviour change, a range of new skills in strategy development, communication research, expertise in IPC and professional management skills for dealing with outside agencies and interaction with PRIs, NGOs and the community, etc., is required. There is thus a mismatch between the skills of the existing IEC staff and the requirements and expectations from them for BCC/IEC.

There was a provision in RCH for hiring consultants who could provide professional support for IEC planning, supervision and management. Unfortunately, it has not been possible to attract good quality professional persons. Retired IEC staff members have sometimes provided some degree of support (as RCH consultants) but by and large the experience in MOHFW and in the states has been disappointing. Hiring professional agencies and assigning work to them has been fraught with difficulties relating to the lack of well-defined terms of reference (TORs) and proper briefing of agencies, mistrust of private agencies and unprofessional supervision of the execution of assignments. If hiring professional agencies is to be encouraged in RCH II, adequate measures will have to be taken to ensure that briefing and supervision is done properly.

The social mobilisation effort through ZSS was an innovative idea in RCH that did not work too well owing to poor understanding and lack of inter-departmental coordination and management. However, obtaining stronger community support for the RCH Programme, achieving better social mobilisation through linkages with NGOs, CBOs and PRIs at the field level, and decentralising the planning of IEC to the district level with adequate flexibility given to the District Health/Medical Officer may be a way out.

IEC/BCC cannot succeed unless its importance is recognised by the entire Health and Family Welfare staff at all levels. It cannot be regarded as the responsibility of the few IEC specialists in the service delivery chain. While they may perform specific technical and management functions, the overall objective of achieving behaviour change through a communication effort has to be a part and parcel of everybody's job. Demand creation and satisfactory utilisation of services is not the objective of IEC alone, but of the RCH Programme itself.

NOTES

1. For a more detailed account of India's Family Planning Communication Programme, see Kakar (1987).
2. The IEC strategy development process described here is taken from ORG (1992).
3. This study titled *People's Perceptions on Population and Development Issues* was conducted in 1994 by Mode Research, New Delhi and a report of the research study submitted to the United Nations Population Fund (UNFPA), New Delhi (see Mode Research 1994).
4. The novel approach for IEC in RCH to address the manifold needs of the 'new paradigm' was suggested in a paper by Cecelia C. Verzosa and Pradeep Kakkar (see Verzosa and Kakkar 1996).
5. This quote is taken from a UNFPA briefing kit on the Cairo Declaration adopted by consensus at the ICPD in Cairo in 1994 on the new approach to reproductive health.
6. This Goal Statement was first articulated in the National Workshop in New Delhi in January 1999 and later refined through the series of regional workshops in Hyderabad, Jaipur and Kolkata and a final meeting in New Delhi in October 2000.
7. The Matrix of Behaviour Change Objectives for RCH was appended to the National Communication Strategy for RCH and was a part of it.
8. During the course of the implementation of the RCH Programme, between 1999 and 2004, I have been a Consultant to the World Bank on IEC and the impressions recorded here are my personal assessment of the RCH communication programme.
9. Several Population Research Centres (PRCs) under (MOHFW) were engaged to conduct evaluation studies of the ZSS scheme operating in the States of Karnataka, Maharashtra, Gujarat, Himachal Pradesh, Assam and Uttar Pradesh in 2000–01. On the basis of the evaluation reports, the scheme was discontinued by the MOHFW.
10. This refers to the Special Scheme for funding IEC activities in 11 weak-performing states under the RCH Programme of the MOHFW, Government of India that was initiated in 2001.
11. This report is based on my visit to Bhubaneswar and adjacent blocks in February 2004.
12. Record of Meeting of State IEC officers convened by MOHFW in October 2000 in New Delhi.

13. Report of National Institute of Design (NID) of the two training programmes in January and July 2000.
14. The figures of persons trained are taken from the report of National Institute of Health and Family Welfare (NIHFW) on progress made with IEC training till 2003.
15. Evaluation Report of Centre For Media Studies submitted to MOHFW in 2004.
16. Shyam Benegal, Amol Palekar and Kalpana Lajmi were some of the filmmakers selected by MOHFW and Amol Palekar's film *Keiri* and Kalpana Lajmi's film *Daman* won awards while Shyam Benegal's *Hari Bhari* received good reviews.
17. While the NPP 2000 gives the figures of relative importance of factors contributing to the continuing population growth, more detailed analyses are to be found in *Beyond Numbers* issue of Seminar (March 2002).

8

Reaching Development to the Rural Poor

USING SATELLITE COMMUNICATION FOR TRAINING GRAM PANCHAYAT MEMBERS IN KARNATAKA

A project of the Indian Space Research Organisation (ISRO), the 1975 Satellite Instructional Television Experiment (SITE) successfully demonstrated the potential of using a satellite-based television system for rural communication. Subsequently, ISRO made further advancement in technology application and demonstrated the potential of satellite-based interactive communication for development, education and training. This interactive 'one-way-video-two-way-audio' system consisted of a head-end studio for live or pre-recorded presentations. The studio was linked to an earth station from where the signals were transmitted to the INSAT (satellite). The satellite relayed the signals for direct reception at field locations with small satellite terminals using the extended C–band antenna, which were then played back through TV monitors located there. The benefit of this system, which provided trainees an opportunity to interact with resource persons and get answers to their queries, was first used in the Open University system spearheaded by the Indira Gandhi National Open University (IGNOU). Later, ISRO piloted several experiments with state governments to extend its application to other areas of training.

Early Experience in 1994-95

In 1994-95, the Government of Karnataka carried out a small pilot experiment using the facility of the ISRO Earth Station in Bangalore. The Department of Women and Child Development (DW&CD), Government of Karnataka, with support from Unicef, developed a set of films for training newly elected women members of the Gram Panchayat (GP) and used the facility offered by ISRO to do an initial training. ISRO provided a limited number of TV sets and satellite receiver terminals located at each district training centre. At that time the challenge was partially to ensure that the hardware performed in a satisfactory manner in the district training centres and there was no break in the talkback facility between the studio and the receiving end. The experts had to be prepared to answer queries simply and directly and the presenter had to ensure that the participants felt comfortable with the interactive medium. The high point of the pilot experiment was the excitement of listening to the other GP members in other district centres raise the same questions or narrate similar experiences. During the interaction the enthusiasm of the trainees was such that an engaging dialogue between GP members in different district training centres took place with the panel of experts in the studio becoming mute spectators in the exchange.

Workshops with GP members were part of an ongoing training programme of DW&CD, Government of Karnataka and UNICEF. Some folk plays were presented (and later filmed and presented) at these workshops to start discussions. These films dealt with the history of Panchayati Raj since Independence and the new provisions of the 73rd Amendment, the reservation of seats for women, learning to function as a GP member, rules and procedures of panchayat meetings, how to ensure transparency and no discrimination, the role of the Gram Sabha and many development issues that were of concern to the GP. Interviews and discussions with GP members were recorded during the workshops. Later the whole series of films was conceptualised and the different elements were put together as a training package.

The 12 films that were produced during that project were made after considerable research and discussion with GP members. These films were developed in a workshop mode; they were not pre-scripted and used no expert lectures. Rather, they reflected the *lived* experience of GP members. Gender and caste equality was a crosscutting theme. The format of the films included a creative mix of workshop discussions, individual interviews and dramatised plays. A large number of NGOs were involved in developing the films. They included Mahila Samakhya, SEARCH, Indian Social Service Trust (ISST), Institute of Development Studies (IDS), Grama Vikasa, Concern for Working Children (CWC), etc. Drama support for the films was received from Rangayana, Chitrakala Parishad and Samudaya. Noted filmmaker Deepa Dhanraj and Navroze Contractor, cinematographer, produced the films.

In this initial experimental satellite communication (SATCOM) training programme, the format chosen for the films mainly comprised the use of folk plays and some discussions. Some of the training workshops were extensively documented on film and later incorporated in the final version of the films. The complete set of films were then used as a supplement to the field-based training conducted by the DW&CD and UNICEF *without* the SATCOM facility and opportunity for participants in different locations to have live interaction with a group of experts who answered queries, clarified doubts and gave more complete information.

The films were very successful with the audience because of their reliance on traditional and easily identifiable folk theatre formats combined with a generous use of songs, music and dance.[1] The narratives were simple and based on real-life experiences of people. Whether it was caste and gender discrimination or corruption and discrimination in the panchayat, the messages came across simply and clearly. Usually the films ended without any resolution of issues or the presenter asking the audience to think about the issues.

The juxtaposition of the interviews with the GP members or the discussions among themselves made it easy for trainees to identify with the issues discussed or observations made in the films. The confidence of these members, their honest admission of difficulties or problems faced within the family and community, the obstacles placed by the powerful upper-caste members, were all real issues faced by the trainees who were mainly dalit women and men. The rich texture of the films provided the key trigger for trainees to find their voice and articulate their feelings, express their ideas and discuss issues openly to arrive at their own resolution. The films helped them to sharpen their focus on issues and clarify their own perceptions.

Lessons from SATCOM for Rural Application

The success of a sophisticated training and communication system based on satellite technology like the SATCOM facility is dependent on the entire system working with precision. Several groups are assembled in different locations. The trainees are often not literate and lack exposure to interactive TV technology. At the same time, they are the principal actors and the entire training system is designed for their knowledge enhancement. Hence, they have to be given space and time to feel comfortable to express themselves, absorb the new information and move to a higher plane of knowledge and understanding. Technology should not be allowed to dictate the training programme with participants being herded to view films and then coaxed to ask questions of a panel of experts.

The use of participatory training methods like games, group work, role play and other participatory activities helped the trainees to feel easy, relaxed and comfortable in the group. The facilitators also prepared the trainees with the protocol of the SATCOM format of TV viewing, discussion and later interaction through 'one-way video-two-way-audio' with the panel of experts. Both elements—the participatory training methodology and the SATCOM interactive TV—together formed the core of the training programme. Needless to say, the logistics of arranging such a large-scale, multi-site training programme were quite forbidding and required careful planning.

It is worth mentioning that several departments of the Government of Karnataka participated in the pilot experiment conducted by ISRO and the Development Education Communication Unit (DECU) in 1994–95. Perhaps the other departments were unsure of the potential of the SATCOM system or their planning and preparation was insufficient. The result was that the materials used by them were more in the nature of government publicity or information dissemination rather than aimed at provoking critical thinking and lively discussion. The firm belief that people can find their own ways to overcome obstacles and solve their problems provided complete information and communication skills were given to them, was lacking. The trainers of the other departmental programmes were not convinced that SATCOM training could make a difference.

The Earth Station of the Government of Karnataka[2]

Spurred by the success of the pilot SATCOM training experiment, the Government of Karnataka invested in an earth station for itself and planned for the installation of TV monitors and satellite receiver terminals at every taluk-level training centre. It must be added that governments in several other states like Madhya Pradesh and Gujarat have also invested in an earth station to link districts and taluks/blocks with the state headquarters. ISRO had demonstrated the efficacy of this interactive satellite-based communication system and, in fact, one of the key objectives of the Indian National Satellite System II (INSAT II) series has been GRAMSAT (Satellite for Rural Communication) or the application of satellite-based communication technology for training and information dissemination to rural areas.

The Abdul Nazir Sab State Institute of Rural Development (ANSSIRD), Mysore, was established as the hub of Government of Karnataka's satellite-based interactive communication centre. The uplink earth station comprises a 7.5-meter antenna and is capable of operating in an extended C–band of the INSAT system. It receives the TV signal along with the audio signal from the studio. These signals are suitably combined, modulated, amplified and transmitted to the INSAT satellite. The studio comprises a studio floor and a studio control room. The teaching end signals originate from the studio floor for transmission via satellite. The studio floor is a TV studio with cameras, VCRs, PCs, tabletop cameras, etc. It is linked to an EPABX capable of queuing three calls from the districts/taluks. It also has a fax line. The studio control room has an equipment console (audio and video) which facilitates the switch-over of the audio and TV signals.

From the studio anchor-persons and panellists facilitate the training through live or pre-recorded presentations, discussions, demonstrations and talkback sessions. These are transmitted to the satellite through the earth station, which is linked to the studio (see Figure 8.1). The satellite relays the TV signals for receptions directly by small satellite terminals and relay through TV monitors located at the training centres in different districts and taluks of the state. The participants at the training centres seek clarifications, raise doubts with the resource persons present at the studio on an audio channel/fax through telecommunication lines located at the training centres. At the studio the questions received from a training centre are looped

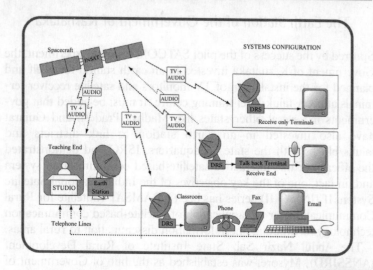

Figure 8.1
ANSSIRD SATCOM System Diagram

back on the audio channel of the TV signal emanating from the studio so that the questions can be heard at all the training centres in different districts and taluks. The response to the question goes on the TV signal and is received by all the training centres.

Since the establishment of a SATCOM training centre in ANSSIRD, the Government of Karnataka has launched several training programmes and the different development departments have used this facility for in-service training of their functionaries. The setting up of the Direct Reception Set (DRS) facility at taluk-level training centres has also progressed and a large number of TV receiver/satellite terminals (also known as DRS) have been positioned in these locations.

Karnataka State Training of GP Members Using SATCOM Technology (2002–04)

The first large-scale and significant training programme initiated by ANSSIRD has been the recent, satellite-based interactive training and communication programme with elected GP representatives. Launched in November 2002, the programme promoted a four-phase, sustained and ongoing interaction among GP members between November 2002 and July 2003. Subsequently, a second stage of the programme in two phases started in October 2003. The first stage covered 18,219 members from 1,318 GPs in 44 taluks of Karnataka. The second stage of the programme, which was launched in October 2003 and comprised two phases, covered 62,000 members from 131 taluks. Probably the largest single training programme for GP members in the country, this training and interaction programme has been of particular significance because of its innovative use of modern communication technology coupled with a participatory training methodology to reach out to people spread over great distances.

The overall objective of the programme was to create a groundswell for democratic decentralisation through peoples' participation in local planning. The success of the earlier training of women GP members that used a set of innovative films to generate confidence among the women members—a confidence that was reflected in the leadership—roles they subsequently played in addressing local development encouraged the ANSSIRD and the Department of Rural Development and Panchayati Raj Institutions (PRIs) to enlarge the scope of the programme to cover most districts and taluks of Karnataka. Developed after months of intensive planning and preparation, the aim of the training was to provoke GP members, especially women and representatives of disadvantaged groups, to take a deeper look at their own situations and collectively struggle to find solutions to their problems. The idea was to also enable them to think constructively in terms of small, practical and viable activities that they could initiate at the GP level, either individually or as collectives. The programme was not envisaged as a one-time event, but as part of a longer process of change and improvement in the PRIs and the entire panchayat administrative system with the active participation of all the stakeholders.

Listed below are some of the components of the training programme, including the satellite-based interactive sessions:

1. The actual satellite-based training for GP members comprised four phases, each of two days duration. Relevant reading materials

were also provided to enable GP members to strengthen their reading and writing skills in order to acquire greater confidence in transacting panchayat functions.
2. During the gap between the different phases, activities were initiated in the villages for awakening the spirit of voluntarism and the desire for self-reliance among the people. For example, contributing labour and time for cleaning village drains, planting trees in schools and anganwadi centres, undertaking school and health mapping, campaigning for compost pits, promoting the use of non-conventional energy sources, etc.
3. Launching a *janadhikara kalajatha* and cultural *mela*s to spread the message of devolution of greater power to the GP (and the Gram Sabha), people's right to information and transparency in government transactions, along with messages on social issues like dowry, alcoholism, wife beating, child/bonded labour, etc.

The training programme consisted of two parts: (*a*) training at the local centres to be facilitated by identified resource persons, and (*b*) through a programme transmitted by satellite from the SATCOM centre at ANSSIRD, Mysore. The objective was to use the latter to strengthen and supplement facilitation by resource persons at the local training centres. It was recognised that the local face-to-face facilitation had to be a strong and necessary component of the training. Indeed, it was a unique feature of this particular programme. Local resource persons were therefore carefully chosen and trained to enable them to discharge their function satistfactorily. During the daily training schedule the two components were intermixed with each other. Consequently, all districts followed a uniform pattern for the activities to be conducted at the district level (Table 8.1).

The entire training approach was based on a few principles that worked as ground rules for the training programme. These were: First, the conviction that GP members were not unaware and ignorant—'empty receptacles' to be filled with information and knowledge. Conversely, that resource persons and officials, whether from the bureaucracy or NGOs, were not the repositories of all knowledge and wisdom. Second, GP members, irrespective of their educational status, came to the training programme with significant prior knowledge, experience and native wisdom. This experience and wisdom needed to be respected and would constitute the base of

Table 8.1
Training Activities

Activity	Agency	Notes
Introduction to the film	Resource persons at the training centres	Resource persons make a brief, introductory statement about the film, without describing the content of the film.
Screening the film	Satellite transmission	Resource persons at the training centres ensure that the TV set is switched on five minutes before satellite transmission for viewing the film.
General responses	Resource persons at the training centres	The first 5–10 minutes after the film is screened will be for any general response that participants have. To carry this further, resource persons invite participants to individually identify three to five of the most significant statements or scenes from the film and share these in small groups. This is followed by a plenary, where more discussion takes place on the themes identified by the participants. In this session, an opportunity is provided for the participants to turn the film over in their mind and allow it to sink in and take root.
Facilitated discussion/ activities	Resource person at the training centres	For each film, the training package includes a set of topics or activities to be facilitated by the resource persons to encourage participants to go beyond the film, carry the issues further, and to think in terms of the steps that they would take in their own circumstances. There are also hints as to how the resource person may conduct various sessions. In addition, there is scope for activities and games in order to ensure that there is no monotony. The variety of activities takes into account various thematic areas as well, e.g., health, education, water and sanitation, etc.

(Continued)

Table 8.1 (Continued)

Activity	Agency	Notes
Question-answer session	Satellite transmission	This session provides an opportunity for participants to explore new ideas/questions, which may have emerged during the general responses and facilitated discussion/activities with a group of panellists at the SATCOM studio in ANSSIRD. During this session the panellists may also take up two or three issues for panel discussion.

further learning. Third, every GP member (Figure 8.2), irrespective of the caste, gender, or community to which he/she belongs, had infinite capacity to learn. No member could be, even inadvertently, scorned or maligned on account of his/her social and economic background. Fourth, mutual learning had to be achieved through a non-hierarchical and informal mode. Therefore, the training programme was designed to ensure participation of all members, not just the few dominant 'eager beavers', through interesting and relevant activities that provided scope for thinking and doing, aroused curiosity and provided the right degree of challenge to find creative solutions to problems. Several devices were used to facilitate such participation—seating arrangements, group formation, techniques for small group and plenary discussions, periodic wake-up activities to break the monotony, and preparing and displaying information and documentation to ensure ownership of the ideas and issues raised during the training.

Finally, as an interactive training programme, the warmth of human contact was very important. The training team was carefully selected to ensure that they were a group of like-minded individuals with a broad outlook on social issues and belief in democratic and secular values and principles. The team had to be objective and non-judgmental, a group with whom trainees could feel free to communicate without fear of ridicule. The training team had to function as facilitators, friends and counsellors who could evoke and stimulate thinking and also accept criticism from the trainees.

Figure 8.2
A GP Member at a Field Centre

The programme was centred on viewing films that were based on the experience of elected GP members, discussing and reflecting on what the films portrayed, and drawing lessons for their own GPs (see Box 8.1 for a brief description of the films).

> **Box 8.1**
> **Films Used for Training**
>
> ### Phase I
>
> *History of the Panchayati Raj*
>
> This dance drama traces the history of the Panchayati Raj from colonial times to the 73rd Constitutional Amendment. It reaffirms faith in Gandhiji's vision of Gram Swaraj focusing on the rationale for decentralisation, power to the people and issues relating to reservation.
>
> *Proxy Elections, Proxy Participation/Voting, Reservations*
>
> Issues of proxy election to GPs and proxy participation in GP meetings. Focus on the rationale for reservation: why, for whom and for how long?
>
> *Learning to Speak*
>
> Focus on illiteracy, caste and gender as impediments to the functioning of GP members. Tampering with the allocation/rotation of seats and reservation for the posts of chairperson/vice-chairperson, which often occurs on account of illiteracy, gender and caste. No-confidence motions, changes in seat reservation on account of illiteracy, gender and caste factors.
>
> *Marching Together—Niyamagalu*
>
> The film deals with rules relating to notice for GP meetings, agenda, ordinary and special meetings, quorum, sitting fees, meeting resolutions, etc.
>
> *Sitamana's Katha*
>
> The film dramatises the anti-arrack movement which grew out of the literacy campaign in Nellore district, Andhra Pradesh. The main message of the film is that when people organise themselves into a collective, they can fight social evils and other forms of discrimination. The film also raises the issue as to whether the GP is intended merely for implementation of 'schemes' conceptualised and designed elsewhere (with perhaps little relevance to the local reality), or whether it has a role in the overall development of the village.
>
> *Song of the Mother Root*
>
> This film deals with the principles underlying the constitution of gram sabhas. The 'Gram Sabha is the Vidhan Sabha of the village' is a powerful statement made by the film.

Box 8.1
(Continued)

Angaivali Aramane

The film *Angaivali Aramane* deals with issues relating to community participation in planning, designing, funding and implementation of development programmes. The film points to the grim and shocking nexus between people with vested interests and how development works are often sanctioned in order to benefit contractors and other people with dishonest motives and corrupt values. It also makes reference to the Peoples' Planning Campaign in Kerala where people took responsibility for identifying their problems and took decisions to find their own solutions.

People's Campaign for Decentralised Planning

This is a simple presentation on the Kerala experience of decentralised planning. It highlights the main features of Kerala's decentralisation, namely: delineation of functions; transfer of human resources; transfer of financial resources; and operationalising decentralised participatory planning.

Namma Panchayati Namma Nirdhare

This film deals with the importance of resource mobilisation by GPs to enable them to finance some of their own development activities, including activities like replacing bulbs for street lights, cleaning drains, etc. The film also deals with issues of property tax collection and points out that the present system of property tax collection is not fair and just.

Property Tax

This short presentation focuses on:

*Anomalies in the present system of property tax fixation and collection,
*Per capita property tax collected,
*Cost of property tax collected,
*How can the system of property tax collection be made more equitable, impartial and just.

Mahiti Astra

Ghost entries in muster rolls of drought relief or other development work gobbling up wages of real residents in the village who are too

(Continued)

> **Box 8.1**
> *(Continued)*
>
> poor to buy themselves two square meals and are without a job. Schoolrooms incomplete in reality but entered in the records as complete. Wells dug only in documents while women fetch water from miles away. Stones never supplied to build a small path bridge, roads never repaired, sums loaned out to the poor for self-employment but embezzled by the rural rich, and so on and so forth—all forming perfect entries in government records. How do the poor know what happens to the minimum wage that is their right? By demanding information contained in official documents. Only after exercising their right to know can the poor strive to fight corruption and demand accountability from those who rule in their name. This is the background to the film *Mahiti Astra* on the right to information.
>
> ### Phase II
>
> *Yaara Shale?*
>
> The focus of the film is on reasons for children not accessing/dropping out of school—seen from the perspective of the community, children and teachers.
>
> *Literacy and continuing education*
>
> Literacy is not merely about learning the mechanics of decoding the alphabet. It is not just about 'reading the word'; it is about 'reading and transforming the world'. Literacy is an empowering skill, which reduces many fears of being lost, cheated and manipulated by others.
>
> *Kula Yaaduvayya*
>
> The film shows community members discussing the existing caste and class problems, which are largely responsible for non-enrolment and the dropping out of SC/ST children, particularly girls.
>
> *Inde Balya*
>
> *Inde Balya* is a documentary that highlights the incidence of child labour in the silk industry. Through interviews with children, parents, policy-makers and development workers, the film draws the attention of viewers to the issue of child labour as a gross violation of children's rights.
>
> *Idu Namma Shale*
>
> This film is a tale of two villages—Hospet and Malkapur. A micro planning exercise had been undertaken in the two villages. The focus

Box 8.1
(Continued)

in the film is on the process through which the community members and teachers sit together to develop an action plan for universalising primary education in their villages. What emerges is the sharp contrast in the relationship between the teachers and the community in the two villages.

GP and Childcare

Gram Panchayat and childcare highlights the objective of the anganwadi programme as well as the problems in its implementation. The film also showcases the experience of a village in Mulbagal taluk of Kolar district where the GP, a local NGO and the DW&CD joined hands in partnership to meet the overall nutritional, immunisation and preschool needs of the children in the village.

Marawwa Maidumbi

The film deals with women's health issues focusing on food taboos imposed on women and the tendency of women to hesitate to talk about their ailments and consequently neglect their health until it is too late. Women also discuss the public health system and the fact that health services are not easily accessible to them. There is also a discussion with ANMs who speak of the problems they face in the discharge of their duties.

RCH Presentation

This film highlights issues relating to Infant Mortality Rate (IMR), Maternal Mortality Rate (MMR), sex ratio, pre- and antenatal care, son preference, etc.

Unheard voices

The film deals with problems faced by people with disabilities and raises issues of their right to live a life of dignity.

HIV/AIDS

This presentation focuses on some of the common misconceptions about HIV/AIDS.

Neerina Sutta

The film deals with the problems of water, water discrimination, quality of water and community involvement in planning and maintenance.

(Continued)

> **Box 8.1**
> **(Continued)**
>
> *Nirmala Gram*
> The focus in this film is on the importance of sanitation and the implications of poor sanitation practices on health and quality of life.
>
> *Hesiru Hejje Patha*
> This film covers issues of conservation with reference to forests, common lands and tanks.

Janadhikara Kalajatha

Kalajatha and folk theatre is a powerful means of communication for community mobilisation. As part of the ongoing training and capacity building programme, a *janadhikara kalajatha* was launched to disseminate the message of 'power to the people'. The *janadhikara kalajatha* was conceptualised with the support of reputed theatre activists with previous experience in conceptualising similar programmes. The Community Mobilisation Unit of ANSSIRD with the support of a Community Mobilisation Consultant implemented the development, planning, execution and supervision of the programme.

The programme of cultural *mela*s and *janadhikara kalajatha*s included:

- Plays and songs on issues relating to decentralisation, people's right to information, transparency in government and panchayat *jamabandi* transactions.
- *Kalajatha*s, which took up social issues relating to universalisation of education and eradication of child labour, social evils such as dowry, the devadasi system and alcoholism.
- Designing and organising public meetings in the form of folk festivals introducing district-specific folk arts—*pattada kunitha*, *viragaase*, *dollu kunitha*, *kamsale* and *nilgar*.
- Organising *prabhat pheri*s, quizzes, poster/drawing competitions, children's *mela*s, collective murals, peoples' walls, *janadhikara mela*s, folk story sessions, puppet theatre, exhibitions, seminars, etc.

The entire programme specifically addressed the problems faced by women, SC/ST people and other marginalised groups. The plays and songs developed for the *kalajatha*s were also filmed and recorded for radio broadcast and TV transmission, in addition to being made available in cassette form.

Voices of the GP Members: Moving Forward

The impact of the SATCOM training for GP members can be best assessed by the articulations of the participants during the training and the actions initiated by them afterwards. The questions and issues raised by them demonstrated their maturity and reflected their sincerity and concern to act as responsible panchayat members. Practical questions regarding the existing arrangements for release of funds, the limited powers of the GP, inequity and unfairness in the devolution of powers, corruption and lack of respect for the GP's concern to protect common property resources within its jurisdiction were quite common during the interactive sessions. The following statements on different issues have been taken from some of the interactive sessions and give credence to the level of participation in the training programme.

Concern for regional imbalance: Isn't the government further aggravating regional imbalances by releasing uniform grants to GPs all over the state, irrespective of their level of development or backwardness?

Questioning domination: Gram Panchayats prepare action plans and submit them for approval to the Taluk Panchayat (TP) and Zila Panchayat (ZP). But the TPs and ZPs quote 'guidelines' and direct us to implement works suggested by them rather than the works identified by us in the action plans! Isn't there a legal provision for GPs to reject the modifications made at higher levels?

Seeking meaningful roles: There are many food security programmes—the Swarnajayanti Gram Swarozgar Yojana (SGSY) school mid-day meals, anganwadi nutrition and drought relief. These are managed by different agencies. Why are they outside the panchayat?

Demanding their right to information: Government releases grants directly to the electricity boards. We have no idea about the payments adjusted to the board! Nor do we have any idea of works taken up from different funding sources by different agencies. Any work taken up in any village of a GP must be implemented only after it is discussed in the Gram Panchayat. Don't people have a right to information about all the works taken up by Gram, Taluk and Zila Panchayats?

Seeking fairness: People in our GP have paid their share towards the Swachagrama and Rajiv Gandhi Drinking Water Schemes—but monies haven't been released and the works have not commenced. How can we ask people to pay property tax? People say finish the work for which we have made contribution, then come back for property tax.

Questioning systems: Does the government deliberately release grants in the month of March so that it becomes impossible for developmental works to be carried out and the funds can conveniently lapse? Why don't funds released by the government reach Gram Panchayats in time? Shouldn't all GP members be informed about the funds released by the government? There is tremendous corruption in the release of funds to the GPs, as officials have to be bribed at every level. Shouldn't funds be released directly by the government to Gram Panchayats?

Demanding rightful share: Why should the Revenue Department collect charges for holding cattle *shandies* (village market) in the jurisdiction of Gram Panchayats? Shouldn't this money come to the GP's share? Why should the Forest Department conduct auctions of the usufructs of tamarind trees? The Revenue Department collects cess, but does not deposit this in the panchayat fund. Is this fair? Muzrai Departments (in charge of temples and religious establishments, etc.) earn profits from temple *jatra*s (fairs). Shouldn't GPs have a share in this?

Suggesting solutions, seeking clarifications: GPs get no income from quarrying. What do we do to raise income from quarries located within our jurisdiction? Poultry farms are raised on small one-acre plots. If we try and tax these, the owners say that poultry farms are agricultural activities under the Land Reforms Act and refuse to pay tax. Is there a solution? People have built structures on government land and agricultural land. How to collect property tax as well as arrears? Poor people pay taxes but the elite bring in party politics and don't pay—it becomes difficult even to initiate attachment processes. How do we handle this? What is the minimum and maximum tax that can be levied on medium and large industries coming under the jurisdiction of Gram Panchayats? Using the tax collected from a ward for improvement of that ward could act as an incentive for people to pay taxes regularly.

Showing concern: *Ashraya* houses are allotted to rural poor through MLAs. *Adhyaksha*s and GP members are more accessible to people, especially poor people. Therefore, the GP should be given powers for house site allotments. *Ashraya* housing requires contribution from beneficiaries, but Indira Awas Yojana (IAY) does not. Why should *ashraya* housing also not be allotted free of cost to people belonging to socially and economically marginalised sections? The beneficiary list identified by Gram Sabhas is routinely changed at higher levels. Isn't it possible to stop this practice? Since the *Ashraya* beneficiaries are originally identified at the Gram Sabha, can't the *Ashraya Samiti* be transferred back to the Gram Panchayat rather than remain with the MLA who heads the committee?

Seeking institutional respect for GP: Banks insist that people bring 'no due certificates' from other banks before sanctioning loans. Why don't banks also insist on 'no dues certificates' from GPs before sanctioning loans? We passed a resolution against the sanction of an arrack shop in our Gram Panchayat. But the Excise Department paid no attention. Taluk and Zila Panchayat engineers do not pay heed to GP members. If GP members are not accorded respect, why should they work as peoples' representatives? Government releases funds to Gram Panchayats along with strict conditions for their utilisation. So what is the point of talking about decentralisation? Decentralisation means releasing funds directly to the Gram Panchayats. But a larger proportion of funds are earmarked for ZP and TP. Is that fair?

Fighting injustice, combating corruption: Why does only 75 per cent of the food grain component reach the godowns? Why is the balance siphoned off? In the food-for-work programme, even if 100 mandays of work are generated, engineers officially permit only 25 mandays. Who will pay for the extra work accomplished? Under SJRY, machines replace the manual labour component, and poor people are deprived of labour/food-for-work. What action is taken against contractors/engineers who indulge in such malpractices? Civil works executed by contractors are often poor. We should make provision for contractors to deposit Rs 5,000–10,000, which can be forfeited in case the quality of work does not meet standards.

Demanding equal status: Why is there no provision for discussion in a meeting convened for 'no-confidence' motions in Gram Panchayats when it is permitted in Parliament? Why are there differences in the tenures of chairmen/vice-chairmen at ZP/TP and GP levels? Why does the reservation roster rotation at the panchayat level not match with the roster rotation at Assembly and Parliament levels?

Protecting common property: We have 20 acres of *gomala* (common grazing ground) land in our panchayat, where we also conduct the weekly *shandy* (market). The *tehsildar* has sanctioned plots to people with political influence for house construction. As a panchayat we objected to the sanction of land, but the *tehsildar* lodged a police complaint against the panchayat. The *tehsildar* and the police have made it impossible for the Gram Panchayat to save the land for *gomal/shandy* use. Panchayats have no control over temple lands. The Department is too far away for day-to-day matters. If these are encroached upon by private parties, who should take responsibility? All revenue sites, land abutting houses, *gomal*s, etc., should be brought under the jurisdiction of the *Gramathana* (village police post), and GPs should have the powers of levying taxes on all such lands so that they are in a position to raise local resources for development.

Seeking answers: Husbands play proxy for women *adhyaksha*s. What is the remedy for this? Gram Panchayat elections are conducted on a party-free basis. Shouldn't the Taluk and Zila Panchayat elections also be party-free?

This list is only illustrative of the countless number of questions that poured in over satellite link from 45 taluks in Karnataka to the SATCOM studio at ANSSIRD in Mysore. Perceptive, insightful and covering a range of issues, including corruption, ethics and governance, right to information, transparency in government transactions, equality for dalits and women's rights, these questions exploded the myth that GP representatives are poorly educated and unable to understand the complexity of self-governance and therefore susceptible to error and mismanagement of their affairs. Consequently, it is usually suggested that external interventions are needed to prepare them and facilitate the development process.

Clearly, as the questions posed during the interactive sessions indicate, there is a demand for genuine, democratic decentralisation through systemic changes that would ensure people's access to information, their involvement in local planning, as also openness and accountability in governmental transactions at all levels. There is an urgent need to debunk the myth of illiteracy, ignorance and incompetence among GP members and encourage the creation of an environment for participatory management. There should be a clear delineation of functions appropriate to each level, linking functions at different tiers with appropriate fiscal devolution. There should be a long-term vision and investment in capacity building.

TOTAL SANITATION CAMPAIGNS: SOCIAL MARKETING IN WEST BENGAL

In April 1999 the Government of India revamped the Central Rural Sanitation Programme (CRSP) and introduced the Total Sanitation Campaign (TSC) to bring about an improvement in the general quality of life in rural areas (RGNDWM 2002). The strategy envisaged to achieve this objective included the following:

- Accelerate sanitation coverage of rural population (up to 50 per cent during the Ninth Plan).
- Stimulate demand for sanitation facilities through awareness creation and health education.
- Cover schools in rural areas with sanitation facilities and promote hygiene education in schools.
- Encourage suitable cost-effective and appropriate technologies in sanitation.
- Endeavour to reduce the incidence of water- and sanitation-related diseases.
- Use IEC to promote sanitation as a means to achieve a better quality of life.

The TSC programme was planned as a community-led and people-centred initiative. A demand-driven approach was adopted with stress on awareness building and meeting the demand with alternate delivery mechanisms. The plan was to gradually phase out subsidies for individual latrine units. The TSC

- offered a broad range of technologies and technology improvisations with reference to customer preferences, construction materials and capacities;
- stressed software, including intensive IEC campaigns in the context of TSC;

- dovetailed the range of funds from GOI and state programmes for rural development;
- fostered the broader participation of NGOs, CBOs and other civil society organisations.

Although 179 districts had been sanctioned a TSC till May 2002, the number crossed 200 soon afterwards. In its early stages of implementation, the latrine aspects of the sanitation effort faced fundamental challenges like low levels of coverage (less than 20 per cent) and low demand. Sanitation reforms required restructuring state engineering departments and moving towards decentralised models of service delivery with the active participation of PRIs, NGOs, CBOs and other civil society organisations. IEC or Communication for Behaviour Change (CBC) had a vital role in bringing about this 'paradigm shift'.

For both sanitation and water, a few states and districts had fully entered the programme effort. Power and funding were being devolved and financial resources had been sent to the districts. Time-bound action plans were drawn up, throwing up the enormous challenges ahead.

Moving from the earlier supply-driven, departmental mode of operation to a more decentralised, participatory and demand-driven mode required a change in the 'mindset' of all concerned—politicians, planners and administrators, the Public Health Engineering Department (PHED) and associated departments, NGOs and other civil society organisations, local communities and individual households. This was a major task, especially since the water and sanitation programme operated in a milieu where development schemes were, by and large, supply-driven. The problem was more acute in the case of sanitation where the present coverage was low and it was also, apparently, given low priority. The role of IEC or CBC was crucial to bring about change in attitude and behaviour.

The West Bengal Rural Sanitation Programme: A Case Study

Unicef and the Department of Panchayat and Rural Development of the Government of West Bengal (GOWB) worked very closely together. The programme strategy for Sanitation and Hygiene

Education was to generate demand through IEC and IPC and offer a range of hardware (latrines) to individual households through the Rural Sanitary Marts (RSMs). The decision to focus on latrines as a priority was a deliberate one, especially since several districts in West Bengal were in TSC. IEC was integrated into the programme strategy, thereby making it a comprehensive marketing strategy.

The firmness of the state government's political commitment to TSC was reflected in the partnership between government administration, elected representatives (state and district level) and other politicians, as well as its collaboration with UNICEF. The State Institute of Panchayat and Rural Development (SIPRD) acted as the Sanitation Cell (presently funded by UNICEF) and was responsible for Human Resource Development (HRD), IEC, Management Information Systems (MISs) and supervision of the district programmes. The materials developed by the Sanitation Cell were focused and appropriate. The quality and content of the TV spots that were being aired and the short-fiction film *Jiban Patua* that was used for motivation were good.[3]

IEC was done in two parts—one week of environment building activities throughout the district using various local/folk media and hoardings, 'miking' or making public announcements through loudspeakers and posters, supported by the mass media (TV spots, etc., initiated by the state government), and later personal contact through house-to-house visits by volunteer motivators or group meetings at the Gram Sansad (electoral booth) level. Social mobilisation was most successful when the mobilisation initiative was internalised by the people and those who had installed latrines became motivators themselves. The village motivator worked at the booth level with one volunteer covering 100–150 households. The village motivators were supported by the GP and RSM. According to a key NGO functionary, on an average it took about five house visits to get an order placed.

In West Bengal the TSC programme operated through the Department of Panchayats and Rural Development (P&RD). In other words, there was no separate set-up and funds were sent to the Zila Panchayat and funnelled through the PRI structure. A Sanitation Cell had been created in the SIPRD in Kalyani to provide technical support for IEC, HRD and MIS. A small district Sanitation Cell had also been created in districts where TSC was operational. However, there was no separate cell for IEC.

The budget for IEC in TSC was a proportion (around 15 per cent) of the budget made available to the district as per the guidelines of TSC. Unicef supported the state-level Sanitation Cell as a carry-over of the earlier Intensive Sanitation Programme (ISP) initiated in the state through UNICEF support. At the district level, the fund for IEC was allocated in a flexible manner. In Malda district for instance, the district was responsible for printing posters, leaflets, stickers (for individual household latrines or IHLs), hoardings and wall writings depending on the requirements of the blocks. The RSMs were responsible for village-level awareness meetings, 'miking' and IPC through the village motivators. On the other hand, in North 24 Parganas, the funds were transferred to the Panchayat Samitis at the block level for further disbursement to the GPs as per the IEC plan prepared by the Water and Sanitation Committees at the GPs and RSMs.

In the case of TSC in West Bengal, a conscious decision had been taken to achieve 100 per cent coverage of IHL construction and the School Sanitation Programme. On the basis of the household surveys done for the district, year-wise plans were drawn up for IHL construction in each block/RSM with appropriate targets for each of the three years of the TSC. Similar targets had been drawn up for the School Sanitation Programme on the basis of a survey of primary schools without toilets and a safe water source.

In consonance with the programme objectives/targets, IEC plans were drawn up to motivate individuals and village communities through house-to-house visits, village meetings and intensive publicity drives using available media.

Inter-sectoral coordination was evident in the case of the School Sanitation Programme where the GOWB had issued orders for pooling together all resources available for school sanitation and water supply and the PRIs/ZPs had been given the responsibility to implement the programme. The role of the Village Education Committees (VECs) was given due importance and they were required to make a token 10 per cent contribution towards the toilet complex (Rs 1,550) in order to receive the balance amount from the TSC funds. Prior surveys by the Education Department ensured that toilets were constructed when water supply had been assured in the school. In other schools, PHED provided the water source from funds available under the Minimum Needs Programme (Eleventh Finance Commission), the District Primary Education Programme (DPEP) and the Accelerated Rural Water Supply Programme (ARWSP). Funds for IEC and HRD were available from TSC, Pilot Projects in Water and Sanitation Sector Reforms (SR) and UNICEF.

There was a regular system of monitoring in place for TSC districts, with monthly monitoring meetings being held at the district level. The *Sabhapati* (Chairman), BDO and RSM represented the blocks. Quarterly meetings for three or four districts (regional) were organised by the Sanitation Cell, SIPRD, followed by an Annual State Review of all districts. At the field level the District Coordinator, who was expected to visit each RSM at least once a month, monitored the activities of the RSM. Apart from these meetings, regular feedback was obtained in the prescribed proforma from the village motivator, GP, RSM and Panchayat Samiti and these were consolidated and sent to the RGNDWM every month. However, this was mainly in terms of physical and financial achievements.

There were several notable features of the West Bengal Rural Sanitation Programme:

- The state government had taken up sanitation as a priority programme and there was full political support for it.
- The earlier experience of the Intensive Sanitation Programme (ISP) and the even prior pilot project in Medinipur district had given West Bengal a head start in designing a programme that was sustainable.
- It was based on the ground reality of life circumstances of the rural poor and therefore the product on offer was the least expensive with a subsidy lower than that offered by GOI.
- The strategy was sound with an aggressive marketing thrust to advocate sanitation and offer a product that had been tested and was affordable.
- The role of IEC and HRD in demand generation was recognised and focused motivation campaigns were launched.
- The key role of RSM with direct commercial benefit as an impetus was a strength of the programme.
- Systems were in place to provide support (IEC, HRD and MIS) at all levels.
- The political leadership and party cadres, the administrative machinery and the NGOs worked together with well-defined roles and responsibilities.

- Even inter-sectoral coordination had been successfully achieved in the case of the School Sanitation Programme.
- All this was made possible because the fund flow was routed through the PRIs that were already well established.
- The close collaboration between UNICEF and the Department of P&RD and SIPRD (Sanitation Cell) was also a strength of the programme.

West Bengal's sanitation programme (see Figure 8.3) stood out as a mature effort that had a clear focus and full commitment of the political leadership and administrative machinery. Building on the past experience, the up-scaled TSC programme planning followed the same pattern of the successful Medinipur pilot project. There was a conscious thrust in West Bengal to focus on the Construction of IHLs and the School Sanitation Programme. The two critical agencies in programme implementation were *a*) the RSMs at block level that were responsible for generating demand *and* ensuring supply of IHLs; and *b*) the State Sanitation Cell, which was responsible for communication (IEC) materials, training support for the District Resource Teams and RSMs, as well as monitoring and evaluation. The materials used for communication were simple and grounded in the life circumstances of beneficiary groups. Likewise, the products on offer were the least expensive with the option of upgradation in due course. The key target group was well defined—those below the poverty line (BPL)—and the challenge was to change their attitude and behaviour towards sanitation generally and open defecation in particular. However, it was the confidence of the programme managers and their resolve to achieve the targets, that was most striking. Hence, the training imparted to village motivators and the IEC materials provided to them were of a high standard and appropriate to the ground situation. It was not surprising that IEC was not regarded as a separate activity but integrated into programme implementation. The judicious use of mass media (TV spots), traditional folk media, posters, question-answer booklets for motivators, prayer books for schoolchildren and handbills on different models of IHL reflected the teamwork that went into the pretesting of the materials for simplicity of language and visuals before arriving at this mix. In short, the political will was evident and the key role

Figure 8.3
Sanitary Latrines and Squatting Platforms

assigned to the State Sanitation Cell and the RSMs was clearly a marker of the programme's success in West Bengal. Both these agencies played a critical role in IEC and demand generation. There is a lot to learn about CBC from the West Bengal sanitation experience.

Box 8.2
Rural Sanitary Marts (RSMs): The Medinipur Story

The date of the wedding was settled and both parties look pleased as the bride-to-be serves tea. She goes up to her father and whispers something in his ear. The father nods and puts a last question: 'Is there a sanitary toilet in your house?' The groom's father and the bridegroom look sheepish and shake their heads. The bride-to-be is pained and disappointed. She can't bring herself to accept the marriage. The father of the groom hastens to give the assurance that the sanitary toilet will be constructed before the wedding.

Thus ends a TV spot put out by the Department of Panchayats and Rural Development, Government of West Bengal. Today, in rural West Bengal, the aspiration of a young bride-to-be married into a household with a sanitary toilet is not at all unrealistic. Thanks to the effort of the Government of West Bengal in collaboration with UNICEF, the unhygienic practice of open defecation is gradually coming down and the toilet is a necessary feature of most households, even relatively poor ones. How did this dramatic change in accepted social behaviour come about?

Since the 1980s there has been an accelerated effort to provide drinking water and sanitation facilities in rural areas. According to latest all-India figures over 87 per cent of rural habitations have access to a safe and clean drinking water source. However, coverage of rural households by sanitary toilets is less than 20 per cent. Consequently, incidence of sickness due to water-borne diseases remains high through contamination of drinking water sources due to open defecation, unsafe disposal of child's stools and poor water-handling practices. Studies showed that loss of person-days at work, loss of schooldays by children and substantial household expenditure on water-borne diseases were common in rural areas.

The initial efforts of the Government to introduce sanitary latrines in rural households during mid-1980s and early 1990s, were characterised by certain limiting factors. The programme did not appropriately address the community's resistance to opt for toilets as an alternative to open defecation. The programme offered relatively expensive toilets with a pucca superstructure—heavily subsidised for the poorer families—and these were evidently out of place in the rural community. The government programme did not offer anything to encourage others to build individual household latrines. In any event, except for the well-to-do, very few had home toilets and open

Box 8.2
(Continued)

defecation was the accepted social practice. There were also space constraints and apprehensions about pollution and bad odour within the home. The toilets supplied to the poorer households did not get used much. The story goes that poor families would use the well-constructed toilet (probably the most secure structure in the home) as a store or a shelter for domestic animals!

The Government of West Bengal, supported by UNICEF, initiated the Intensive Sanitation Programme (ISP) in Medinipur district in 1990. Medinipur is a large district with a population of about eight million (about one-eighth of West Bengal's population) and in 1991, only 4.74 per cent of the population in Medinipur had access to a sanitary toilet.

Working on the premise that there existed a demand for toilets, albeit latent, a demand-responsive strategy was conceived to achieve the objectives. The key communication strategy was to ensure that latent desire for a toilet got converted to an expressed demand for an affordable product. A range of toilet models—from the least expensive ones with just one pit and squatting plate and water seal, to the more expensive ones complete with a brick-lined superstructure and roof—at differential rates was designed to cater to the users. Interestingly, the less expensive toilets could be upgraded at a later stage without much additional cost or difficulty.

The cornerstone of the programme has been the Rural Sanitary Mart (RSM) located at the block level. The RSM is a retail outlet with a sale counter as well as a production centre where toilets are manufactured. It is staffed by two mart managers, two chief motivators with a network of motivators located at the Gram Sansad (electoral booth) level (one for each Gram Sansad or about 100–150 households) and two or three head masons with a group of masons. The sustenance of the RSM depends on its commercial performance. The goal of the RSM is to saturate villages with home toilets. If that could be done, the general health status and overall environmental condition of the villages in the area would improve considerably. The RSM was accessible to all Gram Panchayats in the block. If the households were not fully motivated to adopt a toilet they could be mobilised (through the network of motivators) to adopt a toilet. For every order placed with the RSM, the motivator would get Rs 20. This 'social marketing' approach comprised a crucial element of RSM.

(Continued)

> **Box 8.2**
> (Continued)
>
> The Ramakrishna Mission Lok Shiksha Parishad (RKMLSP), a reputed NGO in the state, was assigned the task of programme implementation. The RKMLSP supports the RSMs at the block level with the help of 16 cluster-level organisations (each managing four to five blocks or RSMs) and 1,000 youth clubs at the village level. The youth clubs, according to their geographical locations, are affiliated to the respective cluster-level organisations and are entrusted with propagating the programme at the grassroots. The coverage of households by sanitary toilets in the district increased to 45 per cent by 2001 (from 4.74 per cent in 1991).
>
> In March 2001 Nandigram II Block of Medinipur District claimed to have achieved full toilet coverage in the entire block. A team from SIPRD conducted a random sample survey (drawn from seven Gram Panchayats) in September 2001 and found that all the 1,741 sample houses visited had sanitary latrines. As a result of the successful implementation of the programme in Nandigram II block, the general cleanliness and overall environmental condition of the villages improved considerably in comparison with neighbouring blocks. A sanitary latrine has become a prized possession of families.
>
> As a direct impact of the programme, since 1999 there has been no incidence of diarrhoeal death in Nandigram II and the incidence of diarrhoea is reducing gradually. While there were 415 cases of diarrhoea reported for outdoor treatment in the block in 1999, the comparable figure had gone down to 264 by 2001. There are 91 primary schools in Nandigram II, which is perhaps the only block where every primary school has a toilet block and tube wells constructed under the School Sanitation Programme. The surroundings of the school are no longer dirty with excreta and infected with the odour of urine. The old picture, the old environment has changed.
>
> In West Bengal's Medinipur district over 400,000 individual household latrines have been built in the last decade and most districts in the state have adopted this very successful model of 'social marketing' based on identifying a felt need and an effective communication effort supported by efficient service delivery.

Source: Ghosh 2003.

ADDRESSING POVERTY: ELICITING PARTICIPATION OF THE RURAL POOR IN MP

The Indira Gandhi Garibi Hatao Yojana or the District Poverty Initiative Programme (DPIP) in Madhya Pradesh was designed to enhance institutional capacity for participatory planning and execution of poverty-reducing interventions. The project attempted to mobilise the poor and the disadvantaged so as to improve their capacity to demand and access public resources. DPIP expected to reduce their poverty by improving their organisation, skills and access to social and economic infrastructure, services and employment opportunities. In order to accomplish this, DPIP attempted to improve the abilities of non-government, government and local self-government institutions to listen to, reach out and serve the poor and disadvantaged clients. Communications and Information therefore had a vital role to play in DPIP.

Some salient features[4] of the project were:

1. The Indira Gandhi Garibi Hatao Yojana is a World Bank-supported project that intended to disburse Rs 6 billion in 14 districts of Madhya Pradesh over a five-year period beginning April 2001.
2. The 14 districts were chosen according to the ranking of the districts in the *State Human Development Report* (HDR) wherein these districts were at the bottom.
3. The institutional structure was decentralised to the village level with the formation of Common Interest Groups (CIGs) in every village covered by the project and decisions were to be taken by the Gram Sabha. Every district had a Process Facilitation Team (PFT) with members drawn from government functionaries as well as NGOs active in the district.
4. DPIP was not rigid about selection of beneficiaries from the BPL category. Since the districts chosen were the poorest according to the HDR, wealth ranking was only used as a measure and the bottom 70 per cent in that ranking became the target beneficiaries for DPIP.
5. DPIP was a demand-driven project and believed that groups with common interest already existed in the villages. The

project, through the PFT, had to mobilise these groups to articulate their demands.
6. The PFT cleared the technical feasibility of the sub-projects articulated by the CIGs up to a value of Rs 300,000 and the administrative sanction was authorised by the Gram Sabha. Beyond that, it went to the district office of DPIP for legal and environmental clearances. The district office could only seek clarifications but not refuse a project.
7. Madhya Pradesh had by then legislated the Gram Swaraj Act under which there were eight Standing Committees that had been constituted for every panchayat and reliance was placed on the Gram Sabha (through its existing or newly constituted Standing Committees).
8. All grant money was transferred to the CIG's bank account in one branch. Fifty per cent was released initially, with the bank releasing the balance amount upon certification by the PFT. This speeded the release of funds and also gave the CIGs a 'valued customer' status since the funds were parked in the bank ahead of requirements.
9. One important objective of DPIP was to get the cooperation of the PRIs and make them its project partners. To this end, it undertook a programme of capacity building for PRIs.

The specific objectives[5] of the Communications and Information component in DPIP were:

- To raise awareness and engage the poor and disenfranchised people to challenge existing barriers and adopt participatory behaviours in the framework provided by DPIP.
- To provide timely and well-targeted information and facilitate learning opportunities for the poor.
- To use peer group and team building exercises to counsel, and use mediation and troubleshooting techniques to reduce existing barriers to change.
- To reinforce people's ability to sustain the new behaviour through positive social support systems.
- To promote a responsive and cooperative environment at the village and district levels that would encourage people to repetitively try and maintain their new behaviour.

Communications in DPIP included all activities connected with reaching information and messages to different audience segments in order to influence attitudes and affect behaviour. It included 'internal' communications addressed to all managers and facilitators working for the project and equipping them with the necessary communications skills to accomplish their goal of changing behaviour among the vulnerable sections of the community. Communications[6] also included all interactive processes and exchange of ideas and experiences among the vulnerable client groups—both as individuals and as collectives—and between them and the other sections of the village community and DPIP project facilitating agents. In order to move from a document that was comprehensive in scope to a delineation of activities, especially for the initial stages of the project, the *key communications tasks* identified for the project were as follows:

1. To make the poor and disempowered individuals, families and groups aware of DPIP (within the broader context of rural development) and the opportunities provided in the project (through their participation in the planning and implementation process) for their economic and social betterment.
2. To establish the credibility and sincerity of DPIP among the poor and marginalised groups so that they took interest in DPIP and came forward to participate in the project by forming CIGs and developed proposals for financing through DPIP.
3. To create an enabling environment within the village community such that the DPIP participatory planning and development process was not resisted or obstructed.
4. To inform other departments and institutions working in the districts about DPIP and obtain their support for the project.
5. To provide a steady stream of information to planners and decision-makers at the state level about the progress in DPIP in order to receive their continued support and make transparent the core values underpinning the DPIP. This flow of information was to be shared with a wider audience through the media and academic discourse on development.

Tasks 1–3 were essentially at the village level and relied on the IPC skills of the PFT with support from a communications team. Local folk media and village performances by travelling troupes, exhibitions and participation in fairs, festivals and village markets, distribution of print materials, group/community meetings and other imaginative and creative participatory learning activities like social resource mapping were the various media available at the village level for these communications activities. Mass media support from radio, television and newspapers was expected to be limited, though interactive radio could play a role in due course during the life of the project. The PFT had to take the responsibility and lead in all village-level activities. However, they required support from a communications team for materials and organisation of shows and performances as and when necessary. In order to discharge this key function, PFT members required intensive training in communications skills development during their induction training for DPIP.

Tasks 4–5 were to be carried out at the district and state levels. Communicating with line departments, banking and other institutions, and eliciting support from the media and the political/administrative establishment were fairly sophisticated and complex tasks requiring significant expertise and planning. These elements were critical to DPIP because without steady support for the project, DPIP could not produce results or achieve its goals. Mass media, media/public relations, orientation workshops and seminars, circular letters (direct mail) and organising field visits for journalists were some of the measures adopted at these levels. While there could be persons designated at the district and state levels to coordinate communications activities for DPIP, they would need to be trained in communications skills as well as in communications planning and management. Outside expertise and assistance from professional agencies was also necessary.

It was clearly and unambiguously understood that an empowerment project had to succeed in facilitating the development of the poor and marginalised groups in the villages. The first three communications tasks at the village level were of prime importance. The other tasks were important in so far as they were necessary for supporting DPIP. Priority for and ensuring the quality of communications at the village level were essential. Establishing trust with the poor and marginalised groups and overcoming opposition from the more powerful sections in the villages was to be effected by the PFT though a process of engagement in dialogue and participation. This was the most critical task in DPIP and it was largely a communications task dependent on the personal abilities and commitment of those engaged in it—the PFT team members. All others associated in the project had the responsibility to make the task less arduous through timely and appropriate support whenever necessary. Establishing rapport with the poor, particularly women, in a government programme is not easy and PFT members had to imbibe the right values and attitudes in order to succeed in their work. This was the focus of the communications effort at the field level.

The other important thing to remember is that DPIP laid a lot of emphasis on 'core values'. These values—participation, empowerment, process orientation, decentralisation, learning, transparency and collaboration—were highlighted in the operations manual as well. The capacity building effort focused in helping the field workers (PFT members) internalise these values, although they were not easy to achieve or transfer.

A Communications Planning workshop[7] was held in Bhopal on 14 September 2001 with the objectives of helping the project team get a proper understanding of the role of communications in a project like DPIP and assisting them to evolve a plan for communications activities. A team of communication experts had made several field trips and interacted with PFT members and DPIP beneficiaries and/or CIG members. The workshop provided state-level DPIP functionaries and participating NGOs an opportunity to get feedback from the communication experts so that they could plan their communication activities better. However, just one workshop would not equip them to do everything by themselves. They would require some help in identifying external resource persons/agencies who could provide support in the future as well. It was expected that these tasks would be attempted on the following day with a smaller group. The report of the workshop, 'To Observe, Listen and Respond', prepared by the Indira Gandhi Garibi Hatao Yojana, highlighted the significant observations made by the experts and their suggestions for the DPIP field team. An attempt is made in this case study to present some of the important issues that came up during the workshop, which may be relevant to similar projects elsewhere.

Engaging in Dialogue with People: A Few Tenets

If a government programme is to engage the poor in a dialogue to address their poverty then it is necessary for the change-agent, in this case the members of the PFT, to first become patient and sympathetic listeners. It was important, therefore, for the PFT to get to know the audience (beneficiaries) and obtain an understanding of their life circumstances. Without establishing this 'trust' no dialogue could be possible with the beneficiaries.

As words are the basic tool of communication, these should be linked or related to (or find meaning/place) in the local milieu or cultural setting. The communication should be attractive, arouse curiosity or interest and satisfy the desire for new information. It should try to replicate the style of the traditional bards/storytellers in villages who lace their narration with humour and local colour.

DPIP was attempting to change behaviour by introducing new ideas and attitudes. Words therefore had to be chosen with care and caution so as not to build up too many expectations. That was the only way to build trust. People are apt to be enthusiastic about government schemes because in the context of *sarkari raj* (government rule) people have learnt to 'praise' the masters in order to get whatever they can from the government: *Jo maalik uski gao* (sing praises of whoever is the master). One had to be careful and not be fooled by this display of enthusiasm. Enslavement of the mind—*gulami ki mansikta*—had to be combated through dialogue and communication. This was not easy but had to be done sincerely.

Applying Communications for Behaviour Change

While there appeared to be a great sense of opportunity in DPIP, there was also a sense of disbelief among the people. With its flexibility and quick, efficient and empowered decision-making, DPIP seemed too good to be true. The team members were enthusiastic, innovative and strongly committed to the DPIP culture and goals. The implications of these were huge and not always easy to absorb or to communicate. Some individuals displayed remarkable stamina for the uncertainties inevitable in their task. Their respect for communities, and particularly for the marginalised, was being rewarded with trust. This was the most essential ingredient for effective communication, not to mention the success of DPIP. The evidence for this was reflected in the group dynamics that was taking place, the willingness to invest from within meagre individual and community resources, and to accept the risks both of traditional suspicions and fears as well as of new responsibilities.

The foundation for successful communication in the DPIP context lay in its capacity for analysis. DPIP had to fight the tendency to think of communication as largely publicity or media products sent from the top to the field. At the same time, the use of communications to change attitudes and behaviours was a concept that the team was still grappling with.

The most important task was that of social and economic transformation through *changing attitudes and behaviours* at several levels. In addition, there was a need for *advocacy* of the DPIP activity and culture among those decision-makers and managers who could sustain it with a supportive environment. *Training and capacity building* would also require communication inputs and skills. *Information and data dissemination* would be a fourth level of communication activity. Exchange of information and experience would be particularly important for pursuing income-generation options. For example, current efforts required a vast range of information inputs on a variety of things—from the cost of musical instruments to prices of competing blankets—as well as on micro-credit sources/procedures. All four communication levels required capacities and clearly assigned responsibilities.

Skill development would need to start with building the capacity for developing communication strategies and plans among the project team. Some would also require proficiency in media skills which could be provided through training by specialist institutions or experts. Access to outside professional services was also important if needs went beyond local capabilities.

Gender Issues: A Communications Priority

Self-confidence (*atma viswas*) was important for women's empowerment. Women wanted water but the men were indifferent to their need. DPIP had to show its concern for women by joining in their struggle and help in meeting their demand for water. PFTs also needed to take proactive steps to respond to the felt needs of women as articulated by them along with their initiative in economic activities. It was necessary to remember that working with women required time and patience since they were not used to expressing themselves and articulating their needs and aspirations. Care would have to be taken to ensure that no pressure was put on women to come up with projects by putting ideas into their heads just because DPIP had to meet time-bound targets for disbursement. If this was not done, the 'process' orientation of DPIP would not be integrated into all aspects of project execution.

Processes completed with men (like the PRA and wealth-ranking exercises) would have to be repeated with women. Otherwise, without adequate information, women would continue to remain ignorant. Given the generally skewed nature of information dissemination, fresh norms had to be established by DPIP in this regard. There were very few women PFT members in DPIP. But if men were to deal with women, the process would take even longer. Meetings with women were necessary if DPIP wanted to focus on women as a primary target group.

The approach that the PFT team had adopted (without formally stating it) was to win the men's trust so that they allowed the team to interact more freely with the women. This approach had a negative impact on the programme's image. The women did not feel any particular ownership towards the project. While their experience of the project had so far been positive and they had seen benefits, these were being mediated to them through men.

DPIP needed to identify key women in villages who could become communicators (acting as a bridge) between DPIP and women. They could take the PFT agenda forward even if they were not members of CIGs. The concept of using women specifically as social mobilisers (as initiated in DPIP) created awkward situations. Women engaged in this process raised many issues, depending on their needs (Box 8.3). PFT only took up income-generation or economic activities. This dichotomy also needed to be resolved.

In addition, it was important to understand that gender was not only about eliciting women's participation but also about addressing the relationships between men and women. It was about changing attitudes. It is striking that though PFT team members frequently described their experience as a process of 'unlearning', none of this unlearning had been with regard to gender!

A discussion with the team on the various types of training and inputs they had received revealed that gender had not been a part of

> **Box 8.3**
> **From the Field Notes of Malini Ghose[8]**
>
> In two of three villages (Umri Devra and Sirpoi) between 20 to 25 women attended the meeting. While women of the common interest groups were present, the meetings soon became much larger and were attended by a number of women. The meetings in these villages were very lively and extended for nearly an hour-and-a-half to two hours.
>
> It was clear from my conversations with the women in the first two villages that such in-depth meetings had not been held with them. In Umri Devra the women were familiar with members of the project team but their interactions with them had been limited. They also said that the *saabs* come regularly to the village to talk to the men.
>
> In Umri Devra, when discussing the sub-projects that women had undertaken (goat rearing) it was evident that they were unaware of many of the details of the project—how much money they had contributed or how much money they had got as a grant. One of the women's husbands had to be called to give us the information.
>
> In the second village we visited the entry point activity had been to deepen a traditional well and to repair a patch of road that becomes unusable in the rainy season and cuts off a section of the village. Here too, women had been excluded from the identification process, details of the finances, etc. Discussions with the women revealed that most of the women present did not think that the activity was beneficial to them. They mentioned that the well that has been deepened is used by the upper castes (and that they were definitely not amongst the poorest in the village).
>
> When asked what activity they would have undertaken, the women said they were not averse to the idea of building a well but it would be in the *bundhel,* a part of the village where no water facilities are available. The village, has a number of wells but there is a serious water problem in that area (which is poorer and inhabited mainly by the lower castes). While we were discussing this with the women we also called in one of the project members to point out that women could, with a little effort, definitely be included in such discussions and need to be given the space to do so.
>
> Two issues struck me while talking of alternatives with women. One, that many of the ideas that women put forward were what one would call 'social projects'. In Sirpoi village during a previous meeting

(Continued)

> **Box 8.3**
> **(Continued)**
>
> with the Gender Coordinator, Shajapur District, women had discussed the option of building a *prasav griha* (clean delivery room). She had mentioned this idea when it was established that due to a lack of space women usually give birth in the *goshala* or some outhouse where the levels of hygiene are obviously poor. Moreover, when there are complications the nearest health facility is faraway and getting there is a problem. The idea had obviously appealed to some women who had discussed this after she had left. But family members had told the women that this was not really an issue and they should concentrate on trying to get the project team to give them a goat-rearing project.
>
> Similarly, in Dhandua village, drinking water is a serious problem. The women discussed this issue at length in their meeting but things reached a stalemate over the 'contributory' payment—which was high—as the group size was small and the estimated costs for this activity high. The women were unable to get their families to commit this amount for a project that would make it easier and less time-consuming for them to fetch water.

these trainings. Even the start-up module, which dealt with basic issues of communication, did not focus on the issue of communicating with women or communicating with a gender perspective.

Mobilising the Rural Poor

The dedication and commitment of the PFT team and their understanding of the underlying principles and 'core values' of DPIP was impressive. However, this understanding was not being converted into an understanding of the people. The experiential understanding of poverty and rural indebtedness among the poor had to be assimilated by the PFT members and DPIP if they were to be successful in their effort.

People were being shepherded to participate in the DPIP project without their experience and understanding being taken into account. What is poverty? What were the poor people's perceptions and experiential understanding of poverty? What was the people's understanding of the village situation? People themselves would have to analyse and understand these issues.

Kalajatha and/or local cultural events were useful tools to raise issues. This was not the same as using traditional/folk performances for publicising government programmes. Scripts were written from the perspective of the poor and their culture was used as an instrument to reflect their own reality back to them in order to deepen their understanding and to help them analyse/reflect on their situation. It was only through such an engagement (in a creative way) with people in their total life circumstances that there could be a meaningful participation of people in DPIP. Unless people's participation took place and they began to own the programme, DPIP would remain a government programme offering limited short-term benefits to some.

There was no 'visioning' taking place in DPIP. People were not thinking (or dreaming) about their future. Unless the PFT members went to the village, no creative thinking would take place. There was no local person to keep the 'DPIP fire' alight. Perhaps something was required to be done about it in the form of a local group of enthusiasts in the village who would act as DPIP motivators/catalysts and support the PFT members on tactical issues and informal discussions. Effort would have to be made to bring up the leadership from the village level to the PFT level. That would ensure DPIP's sustainability.

There would also have to be complete transparency with the people in the DPIP project. For instance, a village bulletin board which gave details about DPIP project funding could be set up so that all the villagers were aware of the facts. People's participation would have to be made DPIP's main agenda.

Envisioning the future would have to be an inclusive process with representation from all sections, including the poorest. The current practice, no doubt adopted in an effort to kick-start the process, was to include only those who could make a contribution. That was alienating DPIP from the poorest. While this was all right as a short-term strategy, it was necessary to correct it as soon as possible. Ways would have to be found to circumvent this problem by imputing value to the labour component so that the poorest could become eligible.

> **Box 8.4**
> **From the Field Notes of K.K. Krishna Kumar**[9]
>
> On 11 September we arrived at Goghari Village at about 2 p.m. It is a small village of around 600 people in 50 households. Most of the people are small and marginal farmers. Soya bean is the main crop. Most of the families own some land. But the village does not have big landowners. There are four or five families who do not own any land at all.
>
> Economic or social hierarchies are not very dominant in Goghari village. All are equally poor. We went around and visited several households. I asked them about the DPIP programme. All of them had heard about it. All of them felt that it was 'some good programme brought by these *saabs* for us'. They felt it was good and may be different. They mentioned that the members of PFT had visited them several times.
>
> I talked to some young men and children in the village. They did not seem to know much about what was going on. They did not seem to think that they could also get involved with it. I was talking to a young man who was involved in tomato cultivation. He spoke to me at length about different aspects of tomato cultivation. There was another person who had ideas about milk and ghee production. I picked up an interesting discussion about food and milk processing with them.
>
> When we were having a discussion with the women's group, I found a majority of them were non-literate. During discussions, all of them showed interest in becoming literate. But no one seems to have thought about literacy in the context of DPIP.
>
> There were ideas and aspirations in the minds of the people. They were talking about problems too—health, education of children. But somehow these did not seem to be feeding into the DPIP programme. People knew about the programme. They even thought it was good but they were not involved fully or feel that they owned the programme. I don't know. I am not sure. But I feel more can be done to involve people in the programme.
>
> I visited Chethpur Gangai on the 12 September. Even though it was very similar to Goghari village geographically, the socio-economic structure was quite different. The landholding varied from four to five acres to 40 acres. The village consisted of two main social groups: Patels and tribals. Each group had its separate areas in the village.

> **Box 8.4**
> *(Continued)*
>
> We spent some time in the house of one of the Patels and discussed generally about agriculture. Soya bean was the main crop. Water was the main problem in this area also. Later we spent a lot of time with the tribal families. They knew about the programme. They also thought it was good. They had faith in the *saabs* (PFT members) who were coordinating the programme. I was so happy to see the commitment of the PFT members. I asked the tribal *dadu* (elder) how they intended to go about planning the sub-projects. He said: 'We will do according to the suggestions of *saab*'. Here again, processes will have to be strengthened to involve the community in depth and make them own it fully and completely.
>
> We participated in the meetings of two self-help groups. The women's group was much more vocal and was discussing the possibility of starting a goat-rearing programme collectively. The men's group was trying to analyse the possibility of initiating a seed bank collectively. Here again, I was able to talk to a number of youngsters and children. They were not involved so far.
>
> Sitting at *dadu*'s place a young adivasi sang a sweet song and played the typical *ektara* (one-stringed musical instrument) that they use. I asked them about songs and dances. They were enthusiastic but did not feel that these had anything to do with the programme.

MANAGING FORESTS IN JHARKHAND: RESOLVING CONFLICT BY ESTABLISHING DIALOGUE

There are roughly 30 million people in India today whose culture is designated as tribal and they are classified in the census records as 'scheduled tribes'. Until very recently, most tribals lived in those less accessible territories that preserved the ecology as it was in classical times. The forests and hills were not subject to the intensive cultivation normal to the overpopulated villages of the plains. However, there has been an organic link between the rural villagers and the tribal people living on the fringes of the forest or within the forests.

For centuries, tribal and rural communities in India had maintained a healthy relationship between nature, society and culture.[10] Indian villages were integrated ecological entities with agricultural fields and common property resources with diverse community controls. Forests were managed and regulated by customary law with specific areas being reserved for hunting by kings and noblemen. Some trees were regarded as sacred and were never cut, and certain areas were considered as sacred groves and nothing was removed from these areas. Even today, such areas in their natural condition are found scattered in different parts of the country.

The commercial exploitation of India's forest wealth started during the British period and the first Indian Forests Act was enacted in 1878. In 1927, the Government of India adopted a more comprehensive Indian Forest Act that emphasised the revenue-earning aspect of forests. Forests were classified into reserve forests, protected forests and village forests. Restrictions were imposed upon the people's rights over the forest—land and produce—in the reserved and protected forests. Local governments were empowered to levy duty on timber and thus forests became a revenue-earning source for the government.

Throughout this period there was enormous pressure on the tribal people and violent conflicts and battles took place with the British government over ownership and control of forest resources. In many parts of the country tribals waged a losing battle to retain access to forests and ownership and control over their ancestral land and burial grounds. Dispossession and alienation from their land resulting in impoverishment have marked the history of the tribal people in India. The problem grew more acute after Independence when the Government of India emphasised the role of forests for national development and the same policy of regarding forest wealth as a source of potential revenue continued as a priority and restricted the access of the tribal people living there in the name of 'protecting' the forests. In the meanwhile protests from tribal communities continued as they struggled for survival in the face of large-scale displacement caused by large dams, power plants and mining activities. It was only in the 1970s, with the increasing concern for environmental degradation and poor rehabilitation record, that the government became sensitive to the issue of livelihood and survival of tribal communities.[11]

Finally, in 1988, the GOI announced a National Forest Policy that acknowledged the presence of forest dwellers and accepted that the rights and concessions enjoyed by them required full protection. In furtherance of this policy, a circular on Joint Forest Management (JFM) was issued on 1 June 1990 based on the successful experiment of joint forestry in West Bengal. This JFM circular from the Ministry of Environment and Forests (MOEF), Government of India, laid down guidelines for giving village communities living close to the forest land usufruct benefits to ensure their participation in the afforestation programme. It was a way of ensuring the cooperation of the local people in halting further degradation of the forests. Thousands of forest protection committees came into being in different states following the JFM circular of 1 June 1990. According to reports, within 10 years of the programme's implementation, over 10 million hectares were brought under JFM. However, there were some lacunae in the JFM circular and to overcome these, MOEF issued a new circular in February 2000 that focused on vulnerable groups like women and the very poor and gave legal backing to people's rights under the JFM and extended it to good forest areas as well. Ironically, there was a simultaneous move to evict 'encroachers' from forests, particularly those that fell within the purview of national parks and sanctuaries established for the protection of wildlife. This ignored the many cases of disputed claims over forest land that arose out of settlement of rights without proper inquiry and documentation and extinguishing of traditional rights of people living there.[12]

Among the more dramatic movements for the right to 'self-rule' by the tribals has been the demand for a separate state of Jharkhand. After a protracted political movement, the state of Jharkhand finally came into being in November 2000 raising expectations of a better deal for the tribal people of this new state. Jharkhand's forests comprise 2.3 mha or approximately 3.6 per cent of India's total forest cover and almost 30 per cent of Jharkhand's total land area. Jharkhand is home to some of the best *sal* (shorea robusta) forests of the country. However, over the years, intense biotic pressure, neglect by people and occasionally problematic regulatory measures have led to the degradation of large tracts of forests in the state. With a population of nearly 25 million, per capita forest area in the state is around 0.1 ha. Almost seven million people or about 27 per cent of the population of Jharkhand are classified by government as belonging to 'scheduled tribes', most of whom live in and depend upon the forest area. Almost 16 million people are estimated to inhabit more than 10,000 villages situated in and around forests. They are dependent on these forests for their wood and energy requirements. A substantial section of the population also depends on timber-based as well as non-timber resources from these forests for livelihood.

The Government of Jharkhand adopted a Joint Forest Management (JFM) Resolution in September 2001 that recognised that 'active participation of the local villagers and public bodies is essential for the conservation and healthy management of the forests'(Government of Jharkhand 2001). The coverage of JFM was expanded to include all protected forests, reserve forests and protected areas of forests. This was also in consonance with the provisions of the Panchayati Raj (Extension of Scheduled Areas) Act of 1996 wherein the Gram Sabha through its elected Gram Panchayat would ensure protection and regeneration of the forest areas within its jurisdiction through the appointment of Village Forest Management and Protection Committee (VFMPC) or Village Eco-Development Committee (VEDC) with regard to Protected Forests (PF) and Reserve Forests (RF), respectively. Unfortunately, in view of the particular situation of Jharkhand, which has a high percentage of tribal population living in abject conditions of poverty (64 per cent of people living below the poverty line), protection of forests was not possible without making adequate provision for livelihood opportunities: otherwise, driven to desperation by poverty, people would enter forests to meet their survival needs.

In order to circumvent this possibility and improve the living conditions of the people, the Jharkhand Forest and Environment Department (JFED) engaged in developing an externally aided project that would 'reduce rural poverty through improved forest eco-system management with community participation'.[13] The overarching principles guiding the project design were the adoption of a locally driven and broadly multi-sectoral approach to enhance livelihoods in forest-fringe villages. The Government of Jharkhand expected this goal to be achieved through the development of partnerships with agencies (public, non-governmental and private) already engaged in integrated natural resources management and community development in the rural areas of Jharkhand. The Government of Jharkhand engaged tribal leaders and other civil society representatives in the development of the project design. This was a welcome step since ownership, access and control over forests—land and produce—have been very contentious issues in Jharkhand. The tribal leadership, in particular, has been very suspicious and wary of the government's initiatives for the development of the state.[14] The Government of Jharkhand was also addressing issues related to development among tribal communities, forest dependent women and other vulnerable groups, and the legal status of VFMPCs and VEDCs through amendments to the State Panchayat Act and the State Forest Act.

In order to establish its sincerity and credibility with the other stakeholders, the JFED designed a communication and consultation strategy for developing the project in a participatory manner. The thrust of the communication effort was to disseminate the state government's seriousness and commitment to regenerate forests in a sustainable manner that would improve the livelihood opportunities of people living in forest areas. The JFED wanted to build a consensus for this vision at the state, district and village levels. The Joint Forest Management (JFM) Resolution (2001) of the Government of Jharkhand was distributed widely and translated into local languages in the tribal areas so that people became familiar with its content. The outcome of this communication strategy was expected to generate a sense of ownership of the project both among the beneficiaries and influential groups; mediate between influential groups that favoured change and groups that opposed it; enable stakeholders, especially community members, to have a direct influence on the project design; and develop partnerships between stakeholders.

The Government of Jharkhand was aware that while its policy framework for the JFM Resolution was progressive, its implementation was at a very early stage. In order to be clear about what worked and how to make things work, the JFED planned to conduct a series of pilot exercises. The objectives of piloting during the project preparation were twofold: *a*) to build partnerships between the JFED and other agencies (public, private and NGOs) to gain experience with new ways of engaging in group formation, micro planning, and micro plan implementation (including forest product marketing, conflict resolution and community procurement), and *b*) to explore options for participatory resources assessment, JFM timber harvesting and working plan revisions (to adequately reflect micro-plan objectives).

JFED wanted to use some well-reputed NGOs and respected tribal leaders (selected by the tribals themselves) to facilitate and steer the consultation process.[15] Existing communication channels like tribal institutional systems (Manki–Munda in Singhbhum district, Para–Rajas in Ranchi district, Majhi–Parganath in Hazaribagh and

Santhal Pargana districts, and so on); NGO/CBO networks; district- and block-level governments; and JFED would be used as a network for the communication and consultation process for project development. Through this process of dialogue, it was expected that a series of pilot activities would be initiated and the identified partners would guide the implementation of the pilots and their monitoring and evaluation. The dialogue began with the objective of raising the level of motivation of communities towards forest protection and integrated approaches to livelihood strategies. Its final objective was to have communities taking full ownership of the forest protection as well as various development schemes and their operation and maintenance.

While JFED was planning a set of communication initiatives to ensure full participation of the tribal people and local communities in the implementation of the JFM, it had to be sensitive to the reality on the ground. For the last several years, in the absence of any effort by the erstwhile Forest Department of undivided Bihar, many self-initiated groups, particularly in tribal villages, had come up with the objective of protecting the forest. Some of these villages had already developed a set of local rules and regulations, with appropriate consideration for the poorest and most vulnerable who depended on selling timber for their survival. Many participants in the brainstorming workshop (2–3 December 2003) believed that the introduction of JFM should take into account these pre-existing local rules for, in many cases, they were well established and accepted after years of implementation.

One of JFM's implementation norms was the development of micro plans. Very often, forest guards and officials developed them with little participation by the people. Maps were not available in the field to show the demarcation of forest areas that fell within the boundaries of the particular village. In the absence of such maps and other information regarding what was possible, local communities felt helpless and accepted whatever the forest department recommended. The micro plan preparation process had to ensure active participation by members of primitive tribes, marginalised ST and SC households, women and women-supported households, those who were physically challenged, the landless, migrants and other people whose survival depended on selling timber. Maps and other information would have to be made available by the JFED and other government departments so that micro planning activities could become meaningful through the active participation of the people. In addition, special provisions would also have to be put in place to ensure that hunters and gatherers had access to protected forests and forest products. Wherever applicable—for instance, where they had temporary settlements—they should be persuaded to take part in the micro-planning activities and targeted services or activities, for instance in terms of adding value to the forest products they usually gathered, should be included in the micro plans.

Most tribal groups, especially in the tribal-dominated areas of the state, had traditional or informal systems of governance, decision-making and conflict resolution that enjoyed widespread legitimacy also among non-tribals. These also worked as effective communication channels and had traditionally dealt with issues related to natural resource management. These channels needed to be recognised and a positive relationship between these traditional informal arrangements and the formal JFM committees needed to be established. This would facilitate the process of group formation and micro planning by giving it more legitimacy and making it more inclusive. Different strategies would have to be piloted for homogeneous and heterogeneous villages. At the same time, conflict resolution mechanisms and communication strategies would also have to be piloted within the context of such informal networks.

The JFED intended to take on the role of a facilitator and provider of integrated technical assistance for furthering the JFM. It also expected to create a relation of trust between the tribals and the JFED. But at that time the JFED did not have the competency or capacity to achieve this end and hoped that other agencies (public, private and NGOs) already engaged in community mobilisation would help it in this task and improve its capacity to handle it departmentally. A process as complex as the issue of building trust and establishing the credibility of a government department with a local community that had been historically adversely affected by the government's development programme was not an easy task. The challenge before the JFED was enormous and the mood of the tribal people was one of deep distrust. The legal issues of ownership and control, the legitimacy and technical competence of the JFED as a protector and conservator of forests, and the genuine interest of the state government to seek true partnership with the people and work sincerely for the improvement of the lives of the rural poor were in serious doubt.

The communication and consultation process that endeavoured to establish a dialogue with all stakeholders and vulnerable groups would thus have to begin with a set of prior conditions and ground rules[16] in order to establish the bona fide intent of the JFED. These conditions would also bind the other partners to engage in the dialogue with an open mind and view the end objective positively. These rules would have to apply to all partnerships in order to ensure that it was a fair deal for all participants/implementing agencies—the forest department, NGOs, CBOs, tribal institutions and local panchayats. Some of the ground rules were:

- *Accountability*: An effort should be made to ensure that the agencies were *downwardly accountable to the communities*. All too often, because of the compulsions of exhausting budgets in time, filing reports and other reasons, the planning process was curtailed and communities were not given a chance to express their choice or participate in decision-making. Such accountability was crucial since it was the communities that had to reap the consequences of the implementing agencies' actions.
- *Transparency and fund flow arrangements*: Participants highlighted the need for transparency and checks and balances at various levels. At the same time, the capacity of communities to access and manage their own funds needed to be increased. It was felt that the upcoming Panchayati Raj elections and the consequent devolution of powers and transfer of funds to communities made the need for capacity building in accounting and finance management a matter of urgency.
- *Inclusion/participation*: Specific and targeted measures and schemes would have to be put into place to ensure that the most marginalised and vulnerable groups (certain STs, SCs, primitive tribes, female-supported households, physically challenged people, the alcohol-addicted, landless and wage labourers, and so on) were actively involved through culturally appropriate participatory approaches and methods in the development of micro plans. Empowerment programmes would be needed to raise their level of self-confidence and awareness.
- *Autonomy of the approval process*: The technical groups that appraised the micro plans would need to be independent from the groups that facilitated the preparation of the proposals. At that time, micro plans were approved by JFED. While

the section dealing with the management of forest resources would necessarily remain under the control of the forest department, one possible piloting activity could be the formation of inter-sectoral teams and tribal leaders monitoring compliance with the rules, and NGOs and other service providers monitoring technical feasibility assessments.

An ambitious programme of capacity building to develop new skills, systems and partnerships for project implementation would have to be undertaken if the project was to be implemented in a satisfactory manner. Training in communication skills was an important element of this. It was also necessary to improve the capacity of the state government through the JFED and other key players to measure forestry programme impacts on poverty reduction and thereby provide a tool for increased transparency, accountability, learning and programme adaptation. Analyses were urgently required for resource assessment, the legal and regulatory framework, marketing, forest management systems and forestry institutions. Without up-to-date information and application of the best of scientific management of forestry and the skill and patience to continue the dialogue with the people living in and affected by forests in Jharkhand, the Participatory Forest Management (PFM) project could fail to meet its overall ambitious development objective.

At the time of writing this case study, the election of a new government in Jharkhand with a slender majority and the likely introduction of the Scheduled Tribes (recognition of Forest Rights) Bill 2005[17] delayed the start of planning the project. However, the process of engaging in a dialogue, tedious and time-consuming though it may be, was the only way to ensure that all primary stakeholders—the tribal leadership and the people living in the forest-fringe villages, particularly the poor and marginalised, as well as the JFED—agreed to work together in a spirit of cooperation and partnership. Giving inalienable rights over forest land to tribals who enjoyed traditional rights over them would have to be supported with financial and other inputs to mitigate the prevailing difficult and impoverished economic circumstances of the people so as to ensure the regeneration of the forest wealth.

NOTES

1. Report of the DW&CD workshops and article by Ammu Joseph (see Joseph).
2. I am grateful to Anita Kaul, who was the Director General of ANSSIRP during the time that the SATCOM system was used for the Gram Panchayat members' training, for giving me access to the video and other materials used for the training programme. She was earlier the Director of the Department of Women and Child Development, Government of Karnataka when the first set of films were used for training women Gram Panchayat members.
3. This case study is based on a field visit by me to several districts of West Bengal in July 2002 in connection with an assignment from Unicef to review the IEC programme.
4. The District Poverty Initiative Programme (DPIP) has been an initiative of the Government of Madhya Pradesh and the features mentioned are detailed in a booklet entitled *Indira Gandhi Garibi Hatao Yojana, Madhya Pradesh* (see Government of Madhya Pradesh 2001).
5. These objectives are taken from the report of the communication workshop for DPIP held in September 2001 under the title *To Observe, Listen and Respond*.
6. The DPIP project report uses the term 'Communications' in a more inclusive sense and in preference over the more usual expression 'Communication for Behaviour Change'.
7. I was a resource person and facilitator for the Communications Planning workshop in Bhopal on 14 September 2001 and helped the DPIP project team to develop some of the next steps. This case study limits itself to that stage of the DPIP since my engagement with the project did not continue thereafter.
8. Malini Ghose was one of the communication and gender experts who had visited the DPIP districts prior to the Bhopal workshop. I am grateful to her for giving me permission to use these extracts from her field notes submitted along with her report.
9. K.K. Krishna Kumar was the communication and social mobilisation expert who visited DPIP districts prior to the Bhopal workshop. I am grateful to him for giving me permission to use these extracts from his field notes submitted along with the report.
10. An inspiring description of this harmonious relationship is found in Richard Lannoy (1971).
11. A detailed historical account of the legislations with regard to forests and the customary rights of traditional tribal communities residing in the forests may be found in a document titled *Natural Resource Legislation and the People*, that was published in November 2000 on behalf of a collective of agencies working for the welfare and rights of forest dwellers.
12. For a more detailed description of the conflict of interest, see Sarin (2002).
13. The description of the proposed Participatory Forest Management Project is based on the Project Concept Note circulated by the Jharkhand Forest and Environment Department at a brainstorming meeting held in Ranchi on 2–3 December 2003.

14. A collective of tribal leadership and supportive activists under the umbrella of Jharkhand Jangal Bachao Andolan have critiqued the JFED project in different forums and made a presentation at a seminar on 'Forest Policy of the Government of Jharkhand: A Civil Society Concern', organised at the Xavier Institute of Social Service, Ranchi, on 17–18 July 2004.
15. The Jharkhand Participatory Forest Management Project Concept Note was shared and discussed at the consultation meeting held in Ranchi on 2–3 December 2003. I was a participant and facilitator at the meeting and my observations are drawn from the discussions that took place at the meeting.
16. It was heartening to note that in spite of the many differences, participants at the workshop were willing to engage in a dialogue and come up with a series of suggestions that could be adopted by the JFED while planning and implementing the pilot project. Some of these are listed in the case study presented here.
17. The Scheduled Tribes (Recognition of Forest Rights) Bill 2005 has proposed that 2.5 hectares of forest land should be given to each nuclear family of tribals (traditional forest dwellers) and the community would have the joint responsibility of forest conservation. The introduction of the Bill ends a long battle between the environmentalists concerned with the depletion of forest resources and threat to the wildlife and the tribal activists agitated over the 'illegal' takeover of their traditional lands by an unsympathetic forest department. For a more comprehensive analysis of the proposed Bill, refer to Sarin (2005).

Conclusion: Communication Challenges in India

Communication for Behaviour Change requires a more professional approach. Over 50 years ago when India began her journey on a path of planned economic development, it was innocently believed that the task of modernising what was believed to be a traditional society was an easy one: economic development with the promise of a better quality of life was sufficient to motivate people to change their beliefs, attitudes and practices. Following the success stories in other developed countries and impressed by the potential of mass media to reach out to people quickly, India adopted a centralised broadcasting model for information dissemination. In the early years after Independence, radio played a crucial role in knitting the country together and fostered a sense of national identity.

However, changing attitudes proved to be a more difficult task. Just as much as providing food security, elementary education and health care, along with employment, to the growing population became a difficult and elusive goal. The top-down model of communication failed to bring about the necessary changes in attitudes and behaviours primarily because there was little appreciation or understanding of the causes of the prevalent behaviours and practices or adoption of adequate measures to address them. A paternalistic bureaucracy claiming to have all the knowledge and authority to deliver social welfare and basic entitlements to people—and failing to do so—further alienated them from the people. So much so that the free services offered by the government in terms of primary schools and primary health care are not trusted and the poor are increasingly beginning to use private facilities.

Because of its failure to deliver on promises, the credibility of the government has been eroded and the basic development paradigm has been questioned. While there has been, over time, some accommodation of alternative approaches to development with increased participation of other stakeholders, this process has been slow.

Apart from that, cultural values and social norms have been very resistant to change, particularly with regard to strongly entrenched attitudes towards women and caste-based social discrimination. Hence, communication effort to change behaviour becomes a complex process and it is unfortunate that, even now, many in positions of power and authority continue to believe that the problem is one of delivering the message to people and merely announcing (broadcasting) new initiatives of the government will solve the problem. There is not enough investment in research to understand the problem, or the patience to recognise that behaviour change is a slow and difficult process requiring a set of initiatives that go well beyond production of media materials with attractive slogans and catch-all phrases. A more professional approach towards communication is necessary if behaviour change is to be brought about through communication initiatives of the programme managers in the social sector.

It is important to note that the private sector and commercial business establishments have successfully used communication tools and methods to generate demand for products and services, even among those groups that are the primary target beneficiaries of development programmes. Perhaps there is a lesson in it for development planners. Marketing of consumer products and services begins with understanding consumer behaviour. Research helps in identifying unmet consumer needs and the objective of marketing is to satisfy those needs at a profit.

The 'marketing model' for generating demand is based on the underlying principle that the perceived benefit to a consumer *must* be greater than the perceived cost. Even a free service, as in the case of primary health care or primary education, have indirect costs in terms of school uniforms, medicines and/or loss of time or wages. The treatment that the poor receive in government health facilities or SC children get from their teachers and peers often discourages users from availing the service as they perceive no real benefit. Very simply, the communication effort has to ensure that the desire of the client for progress, as expressed through a demand for a particular product or service (safe drinking water or primary education) is satisfied by the provision of easily accessible quality service at an affordable cost. The service provider must have the ability to bring this about through matching the demand with supply.

This marketing approach to service delivery is based on the brutal reality of the competitive marketplace where different products and services jostle for attention. While the products and services that the government has to offer in terms of health care, drinking water or basic education may be different from biscuits and tea or soaps and shampoo, the basic human urge for satisfaction of needs is the same. The marketing approach only ensures that the consumer begins to demand the products and services. Therefore, the private sector, with well-established marketing practices, has evolved its communication approach to generate demand through altering consumer behaviour based on research and empirical evidence in the marketplace. There are countless stories of the successful marketing of new products and services—the most recent being mobile telephony—and they are satisfying a consumer need. In the examples cited in this book, we have also seen the successful marketing of individual sanitary latrines in West Bengal that used the same basic principles of marketing in changing attitudes and generating demand for a product.

Apart from other things, the marketing approach to communication makes it possible to measure the effectiveness of communication effort through changes in market behaviour expressed through increased demand for a product or service. If a communication effort does not work, the corporate client would be forced to change the strategy, as the ultimate goal is to increase sales and achieve higher market share for the product. Unlike the social sector, where there appears to be no accountability, the private sector measures its success in terms of the ability of its sales force to deliver clients for its products and services.

In order to be effective, marketers have used research to develop a profile of the consumer and understand their felt needs before devising a communication strategy to address the consumer. The socio-economic profile, educational status, media exposure and language preferences are some of the characteristics that are investigated and the audience is categorised according to different segments, because it is understood that different consumers with their varying backgrounds and preferences would require different approaches and a priority ordering would have to be done to ensure that the most likely categories are targeted first. Research is also used to ascertain consumer preferences for the product and likely barriers or resistances like cultural habits, competition (other products in the market), price, etc. A communication strategy is devised after a complete mapping of the consumer profile and preferences and an understanding of the problems likely to be faced by the particular product or service in the marketplace.

Based on this understanding and identification of different audiences, a creative strategy is designed to communicate the benefits of the product in a manner that will be appealing to the audiences. For this purpose, message design is tested and tried with sample audience groups before the launch of a full-scale, mass media communication campaign. Usually a mix of media is used according to their relative merits and availability for exposure to the target audience. Cost is an important factor that weighs the use of media. For instance, a sophisticated product for a high-income consumer may find favour in the print media because much more can be communicated about the product in print rather that by exposure on television, which is usually the preferred medium for maximum reach or OTS (opportunities to see). The exposure on mass media is also supported with displays at the points of sale and promotional offers to attract consumers. All delivery staff is trained in using the same messages with the target audiences. In a successful business venture, the marketing team works like a well-oiled machine and products are designed and delivered according to a carefully thought out strategy and plan.

This is not to argue that marketers do not 'sell' dreams and seduce the consumer with captivating images that might be misleading. Unscrupulous practices are not the rule, however, and there are regulatory mechanisms and professional associations that lay down a code of ethics and standards that act as safeguards for the consumer. The short point for our purpose in this book is that communication, in order to be successful, has to be based on research and understanding of consumer behaviour, principles of message design in terms of language, easy communicability and emotional appeal, appropriate media choice, training in communication skills, and monitoring and evaluation of the communication strategy effecting mid-course correction if necessary.

The government service providers have, for too long, adopted the attitude that the services, particularly those that are offered free, are a kind of charity or for social 'upliftment'. On the contrary, food security, health care, safe drinking water, shelter and education are basic entitlements committed by the State to all citizens. When the beneficiaries walk away or do not turn up to avail the services on offer, it demonstrates the failure of the State to look after the welfare of its citizens. Indifference and apathy of the government functionaries, lack of information regarding the service on offer and poor quality of service are the main factors responsible for this. Coordination between supply and demand is essential.

Communication works best when it is in tandem with the services on offer. In the commercial world, advertising or promotion of a product or service is usually not undertaken till the supply chain from the factory to the point of sale is ensured and all agents in the chain are well equipped with complete information regarding the particular product or service. There is no point advertising a product that is not available in the shops *unless* it is specifically done to arouse curiosity and not to mislead. In fact, manufacturers would be liable to a backlash from the consumer if the advertised product is not available or the service is below par. While advertising and promotional materials are used to evoke a fantasy in the minds of the consumer, the products specify their features and have to live up to the promise made by the manufacturer or service provider. Failure to do so would immediately be reflected in a drop in sales and market share, apart from loss of reputation. Big companies go to the extent of recalling batches of products if a manufacturing defect is found in some samples as they value the trust consumers place in the particular 'brand'.

Unfortunately, in the social sector there is, more often than not, a break in the supply chain and what is planned and promised is not delivered in terms of the product or services. Take the promise of rural water supply or electricity or primary health care or basic education. Planners and programme managers do their work in New Delhi or in the state capitals with little understanding or patience regarding the situation on the ground. Supplies do not reach on time and often services are not available at the facility. Construction jobs are parcelled out to contractors in bits and pieces so we have the electricity poles but no wires or electricity. Waterpipes are laid but there is no supply of water. The Primary Health Centre (PHC) does not have medicines in stock or the school building is in a pitiable state of disrepair.

It would seem that government functionaries have no respect or consideration for the client because they are poor and disadvantaged. They do not have political clout or influence. Therefore, their misery can be disregarded. Besides, nobody can be made directly accountable since it is always a systemic failure. The functionaries have their own problems as well—of transfers and promotions, reimbursement of travel claims, maintaining registers and other administrative chores. In such a scenario, it is not surprising that people have lost faith in the government, even when it is genuinely interested in their plight. Their helpless and cynical attitude is to grab what little comes their way, especially when a new scheme is announced or the local MLA or MP or other political leaders come to distribute some largesse as a gesture of great magnanimity. The only signal that the poor get from government functionaries is one of authority and power. The feeling that functionaries are appointed to 'serve' the people and that there is a relationship of trust and confidence between them is hardly evident. There are, no doubt, exceptions to this in this vast country but these are too few and far between.

It is not realised sufficiently that any gap between promise and performance affects the credibility of the product or service. This is something that is well understood in the commercial world of business because it is a lesson that has been learnt in the marketplace. The attitude of the government functionary is quite the contrary. Faced with shortage of resources and supplies, the service provider tends to exploit the situation and wields power to dispense favours. Sometimes, frontline workers, who have to face the people, choose to avoid the situation since they are powerless to change the status quo. Those in supervisory positions also feel equally helpless. Frustrated at the poor response to government schemes or insufficient impact (for instance, in the enrolment of children in schools or for immunisation), administrators mistakenly believe that a communication effort or unfocused publicity will generate or boost the demand for services.

It is easier to plan a publicity campaign and spend monies on printing posters and putting out newspaper advertisements and radio and TV spots than it is to ensure that bottlenecks in supplies are eliminated and service providers are in position with a welcoming attitude. As we have seen in the earlier section, the tendency is to 'blame the victim' or hold the poor responsible for their condition. Usually, government publicity does not follow a systematic approach to encourage behaviour change. There is no strategic thinking to ensure that the target groups will try new behaviour. As an NGO activist in the West Bengal Rural Sanitation Programme mentioned: 'It takes five house visits on an average to persuade a family to install an individual household latrine.' Most government programmes begin and end with spending the budget allocated for IEC without bothering about its effectiveness.

The administrative and political leadership do not seriously believe that communication is an integral part of the planning and implementation process. It is regarded as superfluous and a peripheral activity because there is no perceivable impact. For a long time, it was mistakenly believed that the problem is one of management and shortage of resources. The 'supply push' model of expanding services and building infrastructure (more hardware and civil works contracts!) has reached a dead-end with a huge, burgeoning salary bill and no budget for maintenance of facilities. This affects the quality of service and morale of the government functionaries, especially at the field level. Owing to poor management and an indulgent attitude towards government staff, performance is never a criterion for promotion or career advancement. There is a total absence of 'downward accountability' to the people or intended beneficiaries of a government scheme or development programme.

The net result is that people have lost faith in the government for delivery of services. The *mai-baap sarkar* has yielded to a cynicism where the government facility is good for the free benefits it may provide as and when available. It is not the first option for the poor. The local private medical practitioner or even traditional healer is preferred as they are more conveniently available even if they do charge for the treatment. The trend towards moving to private facilities is prevalent not only in underserved areas but also in areas where government services are available. In the primary health care and basic education sectors, two basic entitlements apart from food security, the rude behaviour of functionaries and their absence have forced the people to seek other sources to meet their needs.

In backward districts of the country privately-run primary schools are enrolling children at the cost of poor attendance in government schools. Sometimes, children are registered in government schools to get free uniforms and textbooks (and now perhaps the midday meals), but they attend the private schools for instruction. Government schoolteachers appear to be comfortable with the situation as this means they have to deal with fewer children. Only those who cannot afford to pay for private services send their children to government schools in despair and without much faith in the learning achievement.[1] As soon as the children reach an age where they can earn something through work, they are withdrawn from school. Parents who believe in a better future for their children bear an enormous financial burden and sacrifice their present well-being to afford private schooling and tuition fees. The situation is pretty much the same in health care. Free services are dispensed with so much indifference and lack of care that people would rather pay and get, if not a more competent treatment, at least a more caring response from private doctors.

I have made these general remarks only to highlight the systemic malaise and not condemn the entire government service delivery system. There is no doubt that many government functionaries render dedicated service even in severely adverse circumstances and service conditions. The frontline workers face the brunt of the people's bitterness towards the government and often feel ill equipped to make amends or affect change. With increasing privatisation and general annoyance with the poor record of the government's service delivery, the popular mood, especially among the vocal middle class, is that basic education and primary health care should also be privatised. This is the tragic consequence of the failure of the service provider to deliver services of satisfactory quality. While management of the service delivery needs to be vastly improved and the experience and expertise from the private sector (including the not-for-profit sector) must be sought, the state has to meet its obligation to fulfil the basic human rights and entitlements of every resident—man, woman and child—of this country. Private sources of basic education and primary health care are often exploitative and damaging since they have no regulatory mechanism in place. If people have lost faith in the government service delivery, the urgent need is to restore that faith and trust by ensuring improved service quality and a more sensitive attitude on the part of service providers. Otherwise, in such a context, communication or government publicity will certainly continue to appear to be a waste!

Service providers must be more responsive: people have a right to information. Most people employed in the government are often dissatisfied with their job responsibilities and their service conditions including pay and perquisites. This is particularly true for the social sector with its vast army of health professionals and paramedical staff, teachers (and now para-teachers), educational administrators, village-level workers and others. They are quite unhappy with their employer—usually the state government or local area administrative/ municipal authority. Some unhappiness or frustration may also be due to not being able to perform their duties because of the lack of support in terms of staff and facilities or supplies of consumables like medicines and textbooks. Their most common complaint is about low salary and lack of other benefits and the insensitivity of the authorities to their personal circumstances. There is a general acceptance that merit may not be rewarded and security of tenure cannot be threatened. All this results in poor performance and lack of motivation to do better.

The management of the human development departments (health and nutrition, basic education, drinking water supply and sanitation, etc.) still works on the inherited colonial administrative system based on a culture of patronage and subservience to the authority of 'those above'. Decision-making is centralised and orders are supposed to be implemented by the fieldworkers without question. This discourages participation and/or transparency, with frontline workers in turn exercising the same kind of authority over the beneficiaries with regard to the delivery of services. Insecure themselves and unable to honestly defend the lack of facilities or unsure of support 'from higher authorities', their attitude varies from one of callous indifference to aggressive hostility, taking advantage of the helplessness of the (usually poor and socially disadvantaged) beneficiaries to seek any redress. It is not surprising that corruption prevails in such a system of apparent shortages where only favours are distributed for a consideration.

Most of the field functionaries perceive themselves as performing a specific function (often technical—engineers, doctors, teachers) and their reporting is upward—to their superiors. Contact with people is a necessary evil or an occupational hazard and has to be kept at a bare minimum. Their own authority, reinforced in the minds of the people owing to their non-availability, has to be retained in order to keep the 'system' functioning. The administrative authority in the district headquarters and state capitals have to rely on the reporting from the field to maintain the data flow back to the central authority in distant New Delhi so that funds may continue come in. Monitoring is only in terms of financial inputs and creation of physical assets. Poor quality of service is taken for granted as the 'inefficiency' of the present system. In such a situation it is not easy to attribute the malaise of corruption to individuals or trace the fund flow. Nobody in the government appears to be genuinely concerned about the assessment of benefits accruing to the people in terms of the services provided or improvement in their quality of life. There is also a belief that nothing can be changed and the service providers are doing their utmost under difficult circumstances to reach the services to the people.

In recent years people have become more aware and are now losing patience with the apathy and indifference of the government service providers. The spread of education and the availability of mass media have made people aware of their basic rights and entitlements. As we have seen earlier, the government itself has encouraged voluntary agencies to generate awareness about schemes for the rural poor and supported non-formal education and the creation of a variety of beneficiary groups to take advantage of these development schemes. Slow though it might be, the benefits of development have been spreading and information and services at the local level are more readily available than before. At a price, private facilities are available for primary health care and even basic education, except in very backward or inaccessible pockets.

Nobody is willing to accept poor quality of services any more. The movement for people's right to information and the success of the *Jan Sunwaii* (public hearing) programme, which was so effective in Rajasthan, is spreading to other parts of the country.[2] All documents pertaining to the Rajasthan's government anti-poverty schemes are to be made available on demand and public hearings are conducted at block level where lists or muster rolls of beneficiaries are presented and false records exposed. Many officials have been forced to admit their guilt in falsifying records. Similarly, fraudulent records of construction activities and other misuse of funds have been brought to light by this simple, powerful and effective form of public hearings with village people participating and the media reporting these events. The powerful collusion between officials and local politicians has been shaken by this collective exercise of people's right to know and have access to information.

The government functionary had till then refused to share information and therefore reserved the power to manipulate records under the cloak of confidentiality and 'official secrets'. Once that myth was blown, the officials were exposed and had to take people's views into account. Nowadays, it is not unusual to find details of several government schemes for rural development like the construction of latrines or hand-pumps displayed on village walls so that people are by and large familiar with the size and scope of the project and identified beneficiaries. This is a small beginning on the road to 'downward' public accountability and transparency.

Fraudulent and corrupt practices in disbursement of funds meant for rural development is one measure of the lack of concern and alienation of the government official from the people. The other is the insensitivity of the service provider in health care and basic education. If a teacher does not come to the school for days on end (for whatever reason), it is the poor children who are penalised and prevented from obtaining their basic entitlement guaranteed by the state. Similarly, if the doctor is not available at the PHC or medicines are not supplied regularly, it is the poor patient who is deprived of a service that is supposed to be a welfare measure paid for by the taxes collected by the state. If the ANM does not do her job of registering all pregnancies and conduct the complete schedule of ANC, it is the pregnant women's health that is placed at risk. Very often, people do not have complete information on what services are or should be available to them. They are used to taking whatever little is offered to them. Very rarely do service providers conscientiously offer the complete range of services that are supposed to be delivered to the people.

In recent years, several new measures have been introduced by the government. For instance, the commendable Education Guarantee Scheme (EGS) in Madhya Pradesh gives people in small habitations with 25 children but without a primary school within a distance of 1 kilometre, the right to demand a school provided they are able to identify one person (volunteer teacher or *guruji*). In such a case, the state government is mandated to provide the funds for making a school operational under the management of the village education committee (VEC). Under this scheme, thousands of schools have been opened in Madhya Pradesh and children have benefited from this. Similarly, the Shiksha Karmi scheme in Rajasthan gives the village community the right to appoint a *shiksha karmi* or a volunteer teacher in a school if the regular teacher has been absent for a prolonged period of time or her/his performance has not been satisfactory. These are all small beginnings in the right direction, demonstrating the necessity to be responsive to people's needs and demands.

The government is beginning to recognise that service providers should be made answerable to the people and the community's help and support should be sought to ensure proper delivery of services. The 73rd and 74th Amendment to the Constitution of India established a third tier of governance at the district, Panchayat Samiti (or block/mandal) and panchayat levels, giving PRIs the authority and power to administer a range of development and welfare services locally. However, the devolution of power is not complete and the structure of line departments still derive their authority from the state legislature and secretariat rendering the PRIs relatively impotent. While it is true that the PRIs do not always represent the interests of the poor, reservations for women and SCs/STs ensure the beginnings of participation. We have seen how articulate the women were in the training programmes initiated in Karnataka and how they were able to wrest power and authority from the vested interests that controlled the PRIs.

In Jharkhand, the traditional tribal leadership is questioning the authority of the State Forest Department over forest resources and no progress in regeneration of forests or improvement in the livelihoods of the poor people is possible without engaging people in a dialogue with the government and ensuring their participation in the planning and implementation process. Increasingly, community-based groups and civil society organisations (CSOs) and NGOs are being inducted in the development planning and implementation process though sometimes the relationship is not one of partnership but that of a sub-contractor (alternative delivery mechanism) to whom some work has been assigned or outsourced. This is the old mode of thinking in government where the authority and control is retained with the government and the ensuing relationship is weighed against the local NGO or community-based group.

The state's capacity to deliver good quality services is seemingly stretched beyond repair. There is pressure on government departments to downsize staff and outsource work. While retrenching staff may be very difficult, fresh recruitment is easier to halt. Collaboration and partnership with outside agencies is the need of the hour. Many corporate business houses are willing to fund education and health care projects but are reluctant to put their money into government because of its inefficiency and lack of accountability. On the other hand, they are willing to take on the responsibility of funding and managing activities through NGOs or CSOs. The Pratham[3] experience in primary education is a successful story of partnership between a Mumbai-based NGO that energised the drive for basic education in the city with a single-point agenda of getting 'every child in school and learning'. Some financial (actually more management than financial) support from the corporate sector and cooperation from the

Bombay Municipal Corporation (BMC) enabled Pratham to work with the community through a team of community-based volunteers (young girls trained to assist teachers in schools and parents at home with teaching).

Effective devolution of authority so that decisions may be taken locally at the village or panchayat level (for the management of the primary school or health sub-centre) and collaboration with the community and local agencies in a more transparent mode of partnership and participatory management with control and authority shifting to the community, has to become the new mantra for development activities. The government cannot regain its lost credibility and build trust with the people without showing its sincere intention to involve the community in the development planning and implementation process. Service providers have to change their attitude towards people and alter their way of functioning—a way that only served to reinforce their power and authority without obligating them to render satisfactory service or demonstrate their care and concern for the people. The government, CSOs, NGOs and the people have to come closer together to make this possible.

Changing societal attitudes and norms is crucial. If the service providers in the government social sector consciously assert the authority of their status and position, they only reflect an accepted premise in our society. Subservience to authority is a deep-seated value (*gulami ki mansikta/jo maalik uski gao*) ingrained through centuries of subjugation and exploitation. Such values and norms in society often pose barriers to human development. Indian society is no exception. Discrimination—economic and social—and intolerance are two critical factors that limit progress.

While the Indian Constitution guarantees equal opportunity to all and protects individual life and property, the ground reality is very different even after half a century of India's existence as a democratic sovereign state. Injustice and exploitation exists at every level in our society. Caste, class and gender are the three representations of such social discrimination and oppression. The poor, especially those belonging to the economically and socially disadvantaged sections, are the worst sufferers. More than that, women in such groups are even more vulnerable to exploitation.

Democracy is based on the principle of equality, but social discrimination sustains inequality in society. While the legislations are liberal, social values still adhere to traditional orthodoxy. The practice of child marriage is still prevalent in many parts of the country.[4] The inherent belief and preference for a male heir is held responsible for the ill treatment of girls in the home and misfortune of the young mothers that do not give birth to a son. Discrimination is made in the food and nutrition made available to girls at home and their education is limited as their basic objective in life, it is believed, is to rear children. As they grow older their movement is restricted and childhood ends quickly as girls are prepared for marriage, helping mothers at home and taking care of younger siblings. After marriage it is a life of 'double burden', working to augment family income and doing household chores and bearing the emotional burden of the family. Men continue to enjoy their privileged status and power, even if the circumstances of modern living conditions have eroded the traditional social structure and the comfort of a joint family. This attitude of discrimination towards women percolates all sections of society in lesser or greater measure.

Progressive legislations regarding the rights of women and affirmative action in terms of job reservations are measures that the government has taken to provide equal opportunity for all. However, social norms and attitudes are difficult to modify and those enjoying power and privileges do not give that up easily. The spread of education has been tardy and health care is still skewed unfavourably towards the disadvantaged. It has been observed by commentators that the ruling elite in India has never been serious about ensuring the end of social discrimination and concessions have only been accepted grudgingly. Basic entitlements of food security, health care, education and shelter are not regarded as a social obligation to be ensured but only as a charitable act of the more privileged and powerful.

This underlying attitude of patronage towards those who are socially and economically disadvantaged results in perpetrating a perception of low self-esteem and despair among the poor and disadvantaged. This leads to under-performance and lack of motivation. The same is the case with regard to the ST population who have been traditional settlers in the forest areas of the country. While the Constitution of India guarantees them certain special privileges, the state, in collusion with powerful vested interests, has slowly eaten into those rights, dispossessed them of their ancestral sites and deprived them even of their source of livelihood. Forest and mineral wealth has been exploited for industrial development without much concern for ecological sustainability and, with the loss of forest cover and rapid environmental degradation, the local tribal communities are being further restricted from accessing their natural habitat. Alternative measures like education and health care with adequate opportunities for sustainable economic livelihood have not been provided, leading to further disquiet and alienation from the Indian state.[5]

It is erroneously believed that it is the poor whose attitudes have to change in order to bring about economic and social development. Very often, it is the deep-seated social norms and attitudes that exert influence on decisions regarding what is socially permissible or desirable. Desire for a male heir is so firmly entrenched that the practice of aborting the female foetus (illegally) is spreading rapidly, particularly in the more advanced states. Parents are under pressure to marry their daughters early, even before the legally permitted age of 18 years, and the demand for dowry and harassment of the bride even after marriage continue, often leading to 'dowry deaths'—a euphemism for murder. While it is true that with increasing urbanisation and social mobility values are changing, this change is taking place very slowly. As the pressures mount on those losing their privileges, mostly men of higher social status, they tend to become hostile to the agents of change.

Impatience with the pace of change and intolerance of attitudes and practices of others leads to confrontation. Those in a dominant position—the ruling elite and the organised vocal minority—have asserted their right to plan India's development with little regard or scant respect for the views of the less advantaged and the poor. It is the poor who have been dispossessed by large dams and other industrial projects with very little compensation. The longstanding Narmada Bachao Andolan (NBA) is a testimony to the deep anguish felt by a large number of people at the lack of sensitivity of the political leadership to their plight and the promises of compensation, though much better now owing to media attention, still do not take into account alternative pathways to development based on a more sustainable and harmonious paradigm.

There are countless examples of faulty planning where local views have not been taken into account and development 'aid' has been thrust on the poor—for example, PWD housing and the construction of latrines in a social milieu that may have a different view of design and lifestyle. The result is a waste of resources and criticism of the poor for not appreciating the 'good' that is being done for them. This is surely indicative of a serious lack of understanding and communication on both sides.

The enforced sterilisation campaign in the 1970s to control population growth was based on a limited understanding of long-term demographic trends and inefficient use of natural resources. Today, we have business leaders commenting that our large population (human capital) should be regarded as a potential reserve like oil and should be encashed in the future when lack of human resources in developed countries with an ageing population forces them to turn to India for its abundant supply of 'human capital'.[6] Nurturing this potential source of wealth through better health care and education (that Amartya Sen called 'social opportunity'), with much greater investments in this direction, should be the way forward rather than viewing the growing population as representing a food security crisis with not enough to 'feed the mouths'. This view arises from that fact that those in power fail to see the poor and uneducated as anything other than a 'drag' or burden on their own level of progress and India's ability to achieve standards of living comparable to the West.

In a democracy, such conflict of interest is natural and its resolution can only be obtained through engagement in dialogue and arriving at a consensus. Unwittingly, the government tends to take measures in haste. This has serious long-term consequences. For instance, though India is a signatory to the reproductive health approach to population stabilisation that moves away from population control, many state governments have adopted schemes of incentives and disincentives for birth control. Studies and empirical data show that a large unmet demand for contraception exists, particularly in weak-performing states and provision of information and services would have a positive effect in reduction of pregnancies. Safe male sterilisation (non-scalpel vasectomy) is a measure that has yielded positive results in many districts and has reduced the contraceptive burden on women, fostering a more equal relationship within the family.[7] Instead of encouraging such measures and looking towards improved delivery of health care that may reduce maternal mortality and neonatal mortality, pressing on with population control seems like a shortsighted, authoritarian and retrogressive measure that militates against the basic human rights of individuals in a democracy.

It is true that consultations and consensus building is a tedious and time-consuming exercise. Advocacy and sustained dialogue has to be integrated in the planning process to ensure that later adverse fallouts are avoided. As it is, the government is increasingly realising that managing the vast array of development activities is impossible for it to handle on its own. 'People's participation' and decentralised planning are now the buzzwords of development planning. Apart from the truly panchayat-level planning that has been successfully demonstrated in Kerala, the experience in other parts of the country is not very encouraging. Whether it is health care or rural water supply and sanitation or natural resource management, those responsible for planning (the line departments) do not know how to engage people in the planning process. Government functionaries are used to exercising their authority and people have no reason to believe that things have changed. So, when officials visit the villages armed with their schemes, the people give them answers they want to hear.[8]

In Madhya Pradesh if women are told that they can get loans for rearing goats and therefore should come up with a plan for such activity, they accept this with mild surprise and bewilderment. They may have preferred to opt for some village development scheme that provided drinking water within easier reach or for investment in a 'clean delivery room' in the village to ensure safe delivery rather than for the limited options for economic activities with uncertain future returns. In Jharkhand, forest protection committees are formed to ensure regeneration of the degraded forests under the Joint Forest Management (JFM) scheme of the Government of Jharkhand. Without providing access to basic services like health, education and drinking water, and making sure that the vulnerable poor have means to earn a livelihood, restricting entry to the forest may be counter-productive because the poor among the tribal villagers have used forest produce for their sustenance over generations. Today, if the forests are degraded and logs of wood can be sold only illegally to obtain some income, then they will do that. Planning for regeneration of forests will have to take the vulnerable sections of the tribal community, particularly the women, into the planning process and meet their urgent needs of sustenance of the family. This can be done only on the basis of equality and partnership, by imparting a sense of shared ownership and control over resources.

A shared vision and understanding of development is necessary. India is making rapid progress in many spheres of economic activity, particularly in the information technology (IT) services sector. The economy is more robust and more integrated with global markets and has been registering a healthy rate of economic growth. India's share in world trade is increasing and it is forecast that India will emerge as an economic superpower in the near future. The President of India has been enthusiastic in challenging young Indians to realise India of their dreams. His book (Abdul Kalam, with Rajan 1998), based on the report of the Technology Information and Forecasting and Assessment Council (TIFAC), gave a vision of a developed India by the year 2020. But this vision is not equally shared among all citizens. Even the identity of an 'Indian' is a contentious issue, with a large number of people of Indian origin living abroad staking their claim and right to build the India of their dreams. Many people within India feel alienated from the mainstream as well.

The pull to achieve comparable living standards with the developed countries encourages increased consumption and yesterday's luxuries are becoming necessities today. At the same time, the grinding poverty and squalor evident in our metropolitan cities reflect our inability to provide safe food and water, shelter, health care, basic education and employment opportunities to a large proportion of our population. It would appear that there is more than one consistent image of India that is emerging and each viewpoint is jostling for space and attention.

The rich diversity of our customs and languages, cultures and practices, views and opinions are becoming areas of conflict and tension. The political, economic and cultural domains are battlegrounds for control rather than an arena for cooperation and conviviality. When hunger, disease and lack of education and shelter haunt the lives of millions, the call for building 'world-class' facilities and infrastructure rings hollow. Besides, who defines what is the desirable and sustainable lifestyle for the future? Many advocates in civil society, through the World Social Summit,[9] are forcefully presenting the case for an alternative paradigm for human development that challenges the dominant ideology of consumerism.

The polity in India is fractured and divided with the Centre barely able to hold a tenuous and fragile peace between the states. Fragmentation and injured feelings mark the political rhetoric to capture votes, and the promises made are those of a grand vision of prosperity that is presently highly skewed and spread very thinly. Impatience with the justice system and disappointment with the competitive federal structure has led to an attitude of aggressive, confrontational politics and protests and agitations. The only heartening thought is that democracy has strong roots and with the passage of time, people, especially the downtrodden and dispossessed, are learning to challenge the authority of the state and demanding their basic entitlements.

In this scenario, communication has a crucial role in engaging divergent views in a dialogue, promoting a culture of inclusive politics and development planning, and fostering harmony and sustainable lifestyles. This process has to go beyond the interactive programmes on mass media where those with audience participation rate high with viewers. Fortunately, television is still a medium of entertainment and polls by different channels, though influencing opinions and generating debate (sometimes frivolous or mischievous) do not reflect the realities of India. The last general elections to the Lok Sabha in 2004 proved the TV channels wrong and showed that the enthusiastic urban viewers do not represent the majority of this country. That is a reality that has to permeate all our planning and envisioning of the future.

Differences and diversities exist in every society. India may have a history of caste-based hierarchy, restrictive social customs and exclusion of others from the social structure (though with some degree of accommodation), but the foundations of modern India are based on democratic principles of equality and freedom. Retrogressive tendencies to glorify the rich, our past heritage or romantic projection of an idealised future has to be tempered by the harsh reality of the present times. While there is potential for enormous development and wealth creation and the vast human potential remains untapped, the reluctance of those in positions of power to relinquish control and invite greater participation in governance and development planning inhibits the full realisation of this potential. The first task of creating basic physical infrastructure and provide social security for all citizens still remains to be accomplished.

The rule of law is openly disregarded and violently challenged, very often by the police force itself. The case of the group of women slum-dwellers in Nagpur beating a criminal to death when he was being presented in a court of law because they feared that he would be released without punishment, is a glaring testimony to people's lack of faith in the maintenance of law and order and administration of justice.[10] This is particularly true of the poor and disadvantaged in society. It is openly admitted that the rich and privileged can 'get their way', but those without any 'approach' or clout with the authority cannot hope for justice. This is a sad commentary on the state of affairs in a society of over a billion people where the poor are in a majority.

It is not surprising that conflicts reach flashpoints very soon. All institutions of the state have lost their credibility and the damage to the social fabric is quite severe. At the same time, that there is prosperity and plenitude is clearly evident from the shopping malls and glossy advertisements on colour television. The contradictions in society are becoming increasingly acute. Those who have been kept out of authority and control are not satisfied with the 'crumbs' that are doled out to them in the name of development. Unfortunately, as has been demonstrated in different parts of India, the political masters are unable to make any radical change in the context of a structure that is competitive and reconcile themselves to singing the same tune. Decentralisation of the governance structure with the creation of PRIs showed promise, but that issue is mired in the tussle between the state governments and the PRIs. Similarly, the Centre–state relationship is an equally contentious issue.

Communication is the only way to broaden the agenda of planning and include all stakeholders in dialogue and discussion. It requires patience and humility. Respect for the other viewpoint, trust in one another's willingness and the ability to work together in a spirit of cooperation and harmony are the basic ingredients of a successful model of participatory planning and collective governance. Visioning the future for India has to be a collective enterprise. For many, even to be able to engage in such a process may require the fulfilment of prior conditions of meeting their basic food security and other entitlements. For some others who are excluded on the basis of social exclusion—traditional and new (physically and mentally challenged individuals, people living with HIV/AIDS)—the government will have demonstrate that it cares. It has to be a demonstration of society that 'we care'. Only then can the true vision of India 2020 take shape. This is the biggest communication challenge today.

NOTES

1. The stark reality of the primary education system can be found in a recent publication edited by Vimala Ramachandran (2004).
2. On Wednesday, 11 May 2005, the Indian Parliament (Lok Sabha) passed the Right to Information Bill that makes it possible for people to demand information regarding various government schemes, leading to greater transparency. The public hearings held in Rajasthan spearheaded by the social activist Aruna Roy and other members of Mazdoor Kisaan Shakti Sangatha (MKSS) led to the national campaign for legislation on the Right to Information.
3. Many examples of NGO efforts, including those of Pratham, are documented in a recent publication edited by Vimala Ramachandran (2003).
4. At the time of writing, an anganwadi worker in Madhya Pradesh had her arms chopped off because she tried to educate the community against child marriage and build up public opinion against this practice (see *The Hindu* 2005).
5. The introduction of the Scheduled Tribes (Recognition of Forest Rights) Bill 2005 may be a beginning towards restoring the rights of the traditional tribal people. However, it does not ensure security of livelihood and is also likely to meet with stiff resistance from environmentalists and other vested interests.
6. In his speech at the Platinum Jubilee Annual General Meeting of the Federation of Indian Chambers of Commerce and Industry (FICCI) held on 13 December 2002 in New Delhi, Mukesh D. Ambani made a strong case for greater investments in education and professional skills development in his speech titled 'Unleashing India's True Potential'.
7. For a more complete discussion on population issues see *Seminar* 511 (March 2002).
8. Refer to the observations of K.K. Krishna Kumar and Malini Ghose in the earlier section.
9. The World Social Summit has become the international venue for discussions on alternative pathways to economic development, global governance, human rights, peace and justice. A recent publication edited by Jai Sen and Mayuri Saini (2005) examines the many issues that have been raised at this Summit.
10. In the middle of August 2004, TV news channels and national newspapers carried the story and several groups rallied to the cause of defending the women for taking the law into their own hands.

Bibliography

Abdul Kalam, A.P.J. with Y.S. Rajan (1998) *India 2020: A Vision for the New Millennium*. New Delhi: Viking, Penguin Books.

Agarwal, Binod C. (1978) *Television Comes to Village: An Evaluation of SITE*. Ahmedabad: Space Applications Centre.

Ascroft, Joseph and Sipho Masilela (1989) 'From Top-Down to Co-Equal Communication: Popular Participation in Development Decision-Making'. Paper presented at a seminar on 'Participation: A Key Concept in Communication and Change' held at the University of Poona, Pune.

Atreya, Venkatesh B. and Sheela Rani Chunkath (1996) *Literacy and Empowerment*. New Delhi: Sage Publications.

Banerjee, Sumanta (1979) *Family Planning Communication—A Critique of the Indian Programme*. New Delhi: Radiant Publishers.

Basrai, Farukh (1976) 'The Use of Small-Guage Video for Broadcast in Developing Countries' in *AIBD–CENDIT Manual*. Ahmedabad: Space Applications Centre.

'Beyond Numbers' (2002) *Seminar* 511, March.

Bhasin, Kamala (1984) 'Women's Literacy—Why and How?' Mimeograph.

Bright, Simon (1981) Media for Dialogue and Development: A Study of Strategies for the Participation of Rural People in Production and Reception of Electronic Media. Unpublished MA Dissertation. Reading: University of Reading.

Chakravarty, Sukhamoy (1987) *Development Planning—The Indian Experience*. New Delhi: Oxford University Press.

Chatterji, P.C. (1987) *Broadcasting in India*. New Delhi: Sage Publications (in association with The International Institute of Communications).

da Cunha, Gerson (1993) 'Safeguarding Youth from AIDS'. Paper prepared for UNICEF and the Government of Uganda, November.

Daswani, C.J. (ed.) (2001) *Language Education in Multilingual India*. New Delhi: UNESCO.

Drèze, Jean and Amartya Sen (1995) *India: Economic Development and Social Opportunity*. New Delhi: Oxford University Press.

Frank, Andre G. (1979) *Dependent Accumulation and Underdevelopment*. New York: Monthly Review Press.

Freire, Paulo. (1971) *Education for Critical Consciousness*. New York: Continuum.

———. (1972a) *Cultural Action for Freedom*. London: Penguin Books.

———. (1972b) *Pedagogy of the Oppressed*. London: Penguin Books.

Ghosh, Akhila (1986) 'Demystifying Media with Rural Women' in *Powerful Images: A Woman's Guide to Audiovisual Resources*. Rome: Isis International.

Ghosh, Avik (2003) 'Marketing Sanitation', *The Hindu*, 11 May.

Government of India (1953) *First Five Year Plan*. New Delhi: The Publications Division, Ministry of Information and Broadcasting. January.

Government of India (1986) *New Policy on Education*. New Delhi: Department of Education, Ministry of Human Resource Development.
———. (1994) *Evaluations of Literacy Campaigns in India*. New Delhi: Directorate of Adult Education, Ministry of Human Resource Development.
———.(1992–97) *Eighth Five Year Plan 1992–97*, volume II. New Delhi: Planning Commission.
Government of Jharkhand (2001) Joint Forest Management (JFM) Resolution. September.
Government of Madhya Pradesh (2001) *Indira Gandhi Garibi Hatao Yojana, Madhya Pradesh*. Bhopal: Panchayat and Rural Development Department, Government of Madhya Pradesh.
Hindu, The (2005) 'Social Worker's Arms Chopped Off', 12 May.
Indian Institute of Mass Communication (1993) *Mahila Swasthya Sangathan—An Evaluation Study*. New Delhi: Indian Institute of Mass Communication.
International Institute of Population Studies (1992–93) *National Family Health Survey (1992–93)*. Mumbai: International Institute of Population Studies.
Joseph, Ammu and Kalpana Sharma (1994) *Whose News? The Media and Women's Issues*. New Delhi: Sage Publications.
Kakar, V.N. (1987) *Communication in Family Planning: India's Experience*. New Delhi: National Institute of Health and Family Welfare.
Kapoor, Sushma and Anuradha Kapoor (1986) *Women and Media Development: A Report of the South Asian Regional Workshop*. New Delhi: CENDIT.
Kapoor, Sushma and Namita Unnikrishnan (1990) *The Woman as Learner—Not Just Another Statistic*. New Delhi: UNICEF.
Karnik, Kiran (1976) 'A Systems Approach', *Seminar*, 232.
Karnik, Kiran and B.S. Bhatia (1985) The Kheda Communications Project. Unpublished Paper. Ahmedabad: Development and Communication Unit (DECU), ISRO.
Kothari, Rajni (1989) *Politics and the People: In Search of a Humane India*, volumes 1 & 2. New Delhi: Ajanta Publishers.
Kumar, Krishna (1982) 'Politics of Literacy', *Adult Education and Development*, No. 19, September.
Lannoy, Richard (1971) *The Speaking Tree*. New Delhi: Oxford University Press.
Lasswell, H.D. (1948) 'The Structure and Function of Communication in Society' in L. Bryson (ed.), *The Communication of Ideas*. New York: Harper and Brothers.
Lerner, Daniel (1958) *The Passing of Traditional Society: Modernising the Middle East*. New York: The Free Press.
Low, Colin (1976) 'The Film as a Social Mirror' in Andreas Fuglesang (ed.), *Filmmaking in Developing Countries: The Uppsala Workshop*. Uppsala: Dag Hammarskjöld Foundation.
MARG (1992) Assessment of the Mass Media Campaign for NLM. New Delhi: Marg.
McKee, Neill (1992) *Social Mobilization and Social Marketing in Developing Countries: Lessons for Communicators*. Penang: Southbound.
McLuhan, Marshall (1969) *The Gutenberg Galaxy: The Making of Typographic Man*. New York: Mentor (first published 1962).
Melkote, Srinivas R. (1991) *Communication for Development in the Third World—Theory and Practice*. New Delhi: Sage Publications.

Mode Research (1994) *Perceptions on Population Development Issues*. New Delhi: UNFPA.

———. (1995) *Attitudes Study on Elementary Education in India*. New Delhi: UNICEF.

MOHFW (2000) *National Communication Strategy for the Reproductive and Child Health Programme*. New Delhi: Ministry of Health and Family Welfare, Government of India.

National Population Policy (2000) New Delhi: Department of Family Welfare, MOHFW, Government of India.

Operations Research Group (1992) *IEC Strategy for Health and Family Welfare* (draft report). New Delhi: ORG (with technical assistance from Johns Hopkins University).

Page, David and William Crawley (2001). *Satellites over South Asia*. New Delhi: Sage Publications.

RGNDWM (2002) *Guidelines on the Central Rural Sanitation Programme (CRSP)/Total Sanitation Campaign (TSC)*. New Delhi: Rajiv Gandhi National Drinking Water Mission, Department of Drinking Water Supply Ministry of Rural Development, Government of India. May 2002.

Ramachandran, Vimala (2003) *Getting Children Back to School: Case Studies in Primary Education*. New Delhi: Sage Publications.

———. (2004) *Gender and Social Equity in Primary Education: Hierarchies of Access*. New Delhi: Sage Publications.

Rao, Y.V. Lakshmana (1966) *Communication and Development: A Study of Two Indian Villages*. Minneapolis: University of Minnesota Press.

Rogers, Everett M. (1962) *Diffusion of Innovations*. New York: Free Press.

Sarabhai, Vikram (1969) 'Television for Development'. Paper presented at The Society for International Development Conference, New Delhi.

Sarin, Madhu (2002) 'Comment: Who is Encroaching on Whose Land?', *Seminar* 519, November.

———. (2005) 'Scheduled Tribes Bill 2005: A Comment', *Economic and Political Weekly*, 21 May.

Schramm, Wilbur (1964) *Mass Media and National Development*. California: Stanford University Press.

Schultz, Charles (1973) Communication and Social Change: Video Tape Recording as a Tool for Development. Unpublished mimeograph, Food and Agriculture Organisation (FAO).

Schumacher, E.F. (1973) *Small is Beautiful: Economies as if People Really Mattered*. New York: Harper and Row.

Sen, Jai and Mayuri Saini (2005) *Are Other Worlds Possible? Talking New Politics*. New Delhi: Zubaan.

Singhal, Arvind and Everett M. Rogers (1999) *Entertainment–education: A Communication Strategy for Social Change*. Mahwah, New Jersey: Lawrence Erlbaum Associates.

Tharakan, Michael P.K. (1990) The Ernakulam Total Literacy Programme: Report of the Evaluation. Trivandrum: Centre for Development Studies. Mimeograph.

Verzosa, Cecelia C. and Pradeep Kakkar (1996) 'Information, Education and Communication Efforts and Social Marketing' in *Supplement to India's Family Welfare Programme—Moving to an RCH Approach*. Washington D.C.: The World Bank.

Index

Aajkaal, a Bengali newspaper group, 143
Accelerated Rural Water Supply Programme (ARWSP), 225
Action India, 61–62
adult education centre (AEC), 91, 97, 100, 104; day and night, 115
adult education programme, existing, 114
adult, exercise of, franchise, 70
advertising, and promotional materials, 259; NLM's campaign, 140
advocacy: 72, 271; audience for, 72; for gender equality and equity, 161; interventions based on normative research, 169
Akashvani Pathmala, 104
All India Radio (AIR), 31, 103, 121; government-controlled, 55; producers, 106; relay stations, 107
anganwadi, 164; centres, 208; workers, 167
Abdul Nazir Sab State Institute of Rural Development (ANSSIRD): 205, 207; community mobilisation unit of, 216; establishment of a SATCOM training centre in, 206
Antenatal Registration and Care (ANC), 187, 265
arivoli (literacy), activists, 131; plays and songs, 132; troupe, 132
Ashraya, beneficiaries, 219; housing, 219; Samiti, 219
Astha, Bombay, 61
audio cassette recorders, 56
audio tape recorders, full-track, 55
Aurat ki kahani (the story of a woman), 93, 96
Auxiliary Nurse/Midwife (ANM), 265

awareness, generation, 157; of human rights, 70; public, campaign, 86

BASNEF (Behaviour, Attitudes, Social Norms and Enabling Factors) model, 74, 76; of communication for behaviour change, 78
Bharat Gyan Vigyan Jatha (BGVJ), 73, 83, 133–34, 137–38; organisers, 138
Bharat Gyan Vigyan Samiti (BGVS), 102; volunteers of, 136
Block Extension Educators (BEEs), 170
Bombay Municipal Corporation (BMC), 267
broadcast, engineers, 44, 54; of war, 25
broadcasting, 29, 55; alternative to, 49; and film production, 148; commercial (Vividh Bharati), 56; contribution of the, system to national development, 51; expansion of the, systems in developing countries, 27; rural, 31; traditional, model, 50

campaign, creative strategy for, 122
campesinos, 51–52; semi-literate, 52
C-band: antenna, 201; of the INSAT system, 205
CENCIRA (translated as National Centre for Land Reform Education and Research), 51
Central Electronic Engineering Research Institute (CEERI), 107
Central Institute of Education Technology (CIET), 114
Centre for Development of Instructional Technology (CENDIT), 56–57, 61–62, 92

Centre for Media Studies (CMS), 192
Centre for Science and Environment (CSE), 64
Centre of Development Communication, 56
Centre–state relationship, 274
CESPAC (translated as Audiovisual Teaching Services Training Centre), 50–52
Chalo Padhayen, Kuch Kar Dikhayen (teach someone, do something worthwhile), 119, 121, 140
change, agents of social, 26; behaviour, 74, 165, 169, 171; behaviour, effort, 76; social, 46, 136
Chauraha, 111; films and print materials, 114; lesson from the, experience, 113; storyline of, 112
CHETNA, 192
Child Marriage Restraint Act of 1978, 196
Child Survival and Safe Motherhood (CSSM), 166
child, delaying the first, after marriage, 160
childbearing, repeated, 60, 92
childbirth, 60; risks during, 174
childcare, discriminatory, practices, 196
Chipko movement in the Garhwal Himalayas, 64
cinema, 25; popularity of the Hindi, and film music, 43; vans, mobile, 34
civil society organisations (CSOs), 266–67
collaboration, and partnership, 266; between PRIs and NGOs, 20
colour television, introduction of, 69
commercial studios, 56
communication, 20, 44, 46, 71–72, 74, 173; activities, 186; and human development, 69; and its role in development, 25; application of satellite-based, technology, 205; application of, technology in development, 17; application of, technology in social and human development programmes, 21; application of, technology in the social sector, 18; applications of, technology in development, 17; bottom-up mode of, 64; challenges, 172, 255; channels, 47, 152, 153, 164; components of, strategy, 74; coordinated, activities, 73; effort to change behaviour, 256; evaluation of the, strategy, 258; existing, channels, 247; face-to-face, 133, 157; for behaviour change, 222–23; for behaviour change, approach, 73, 76, 78, 255; for development, 73; framework of, for human development, 68; importance of mass, media, 25; importance of, in facilitating social change, 21; innovative applications of, technology, 18; inter-institutional, 153; lack of, infrastructure, 33; marketing approach to, 257; materials, 87; media, 83; model of, for development, 72; modern, technology, 207; new approach of, for behaviour change, 77; new approach to, for development, 71; NLM's, campaign, 122; planning, 194; process, 71; professional approach towards, 256; programmes, 18; research-based approach to, 71; role of, in development, 30; satellite-based interactive, for development education and training, 201; skill to improve group, 157; small group, media, 44; state-specific, campaigns, 172; strategies, 157, 258; strategy and its implementation, 17; strategy for NLM, 115; supplementary, materials, 87; support, 86; technology, 18; top-down, model, 42, 255; use of, for literacy and empowerment, 81
community listening sets, 31
Community Needs Assessment (CNA), 19, 167

INDEX

Community-based Organisations (CBOs), 170
Congress government, 65, 121; party, 33, 37, 65
consensus building, social and political, 18
Constitution of India, 73rd and 74th Amendments to, 18, 265
consumer, awareness, 70; goods, marketing of, 70; products, branded, 70
contraception, 270; demand for temporary methods of, 160
contraceptive use, 164
control over resources, 271
corrupt practices, fraudulent and, in disbursement of funds, 265
crafts, traditional, 40
credibility, of the government programme, 82; of the public health system, 168
criticisms, specific, of government programmes, 50
cultural *mela*s, 208, 216

dai, elderly, 94, 98; trained, 94
deaths, starvation, 102
decentralisation, 30, 219
Delhi Doordarshan Kendra (DDK), 100
democracy, 25, 267, 270
Department of Family Planning, 35, 147
Department of Panchayats and Rural Development (P&RD), 223, 226
Department of Rural Development and Panchayati Raj Institutions (PRIs), 207
Department of Women and Child Development (DW&CD), 202
developing countries, 41; economic development of, 26
Development Education Communication Unit (DECU), 104, 107, 204
development, alternative paradigm of, 43; agenda setting process, 18; barriers to human, 267; benefits of the, programmes, 69; economic and social, 36; economic, of European and North American countries, 40; management of the human, departments, 263; messages, 43; new, paradigm, 84; of capitalism, 40; paradigm, 40; paradigm pursued in India, 69; planning, 273; progress and, 25; rural, 20; social, and population control, 150; workers, 45
Direct Reception Set (DRS), 206; for TV viewing, 38, 52
Directorate of Adult Education (DAE), 88, 91
Directorate of Field Publicity, 34
discrimination, 267; against the girl, 94; against the girl child, 60; against women, 89, 90; caste-based social, 256; gender-based, 89
District Adult Education Officers (DAEOs), 107
District Health Officers (DHOs), 183
district health staff, 183
District MEIO (Mass Education and Information Officer), 188
District Primary Education Programme (DPEP), 224
documentation (written and audiovisual), 87; audiovisual, 57
Dohra Bojh (Double Burden), 61–62
donor agencies, external, 194
Doordarshan: 57, 92, 121; revenue for AIR and, 121
dowry, deaths, 59, 269; demand for, 269
District Poverty Initiative Programme (DPIP), 234–38; advocacy of the, activity, 237; credibility and sincerity of, 233; project funding, 241; success of, 236

each-one-teach-one programme, 118
early childhood care and education (ECCE), 124
education, and health care, 269; for all, 82–83; for girls, 160
Education for Women's Equality (EWE), 89

Education Guarantee Scheme (EGS), 265
education, 270; access to formal, 33; adoption of the extension, approach, 35; adult, classes, 97; adult, functionaries, 120; adult, instructors, 100, 104; adult, programme, 117; banking concept of, 45; basic, 262; expansion of, and health facilities, 34; formal, 33; informal, 45; spread of, 268; women's, 114
educator, adult, 45
elite, ruling, 40, 46
Emergency, imposition of, 65; period (1975–76), 37, 111, 148
Empowered Action Group, 186
empowerment, women's, 196
enter-educate, concept of, 156
environment, devastation of the, 59; movement, 59, 65, 70
Ernakulam (and later Kerala) model of social mobilisation, 108
Ernakulam campaign, success of, 126
Ernakulam district literacy campaign, 125–26; achievement of, 139
European Commission (EC), 189
Expanded/Universal Programme of Immunisation (EPI/UPI), 67

family planning, adoption of, methods/techniques, 149, 153; communication programme, 35, 148; programme, 34, 147–48, 164
Federation of Indian Chambers of Commerce and Industry (FICCI), 187
female foetus, practice of aborting, 269
Film Division, 55; documentaries, 34
film, 35 mm, production equipment, 55; and audiovisual materials, 34; centralised production of, 36; documentary, production, 55; documenting the experience of the TLC campaigns, 123; exhibition of, 36; field recording on 16mm, 57; information, 34; information through, and radio programmes, 35; messages promoted through the, 36; on government programmes, 34; publicity, 32
filmmakers, independent, 55
First Citizen's Report on the state of India's Environment, 64
first UN World Conference on Women, 59
foeticide, female, 172, 195
folk drama performances, 185
folk media, 83, 136, 156–57; campaign, 157; forms, 42; issue of, as an appropriate tool of communication, 137; local, 171, 194; use of, in a social communication; process, 136
Food and Agriculture Organisation (FAO), 44, 57
forest, and mineral wealth, 268; commercial exploitation of India's, wealth, 244; demarcation of, areas, 248; dwellers, 244; in Jharkhand, 243; loss of, cover, 20; protected, 146; protection committees, 245, 271; reserve, 146
freedom of the press, 55
Freire, Paulo, 45, 49, 86, 90

Gandhi, Mahatma, 44
Gandhi, Rajiv, 81, 88; government, 66, 114
garibi hatao, 65
gender, and caste equality, 203; caste, class and, 267; equality, 161; equality and equity, 161; injustice, 195; issues, 237
Government of India Act of 1935, 29
government service providers, 258, 264
Gram Sansad (electoral booth), 223
Grama Vikasa, 203
GRAMSAT (Satellite for Rural Communication), 205
Green Revolution, 32, 34–35; beneficiaries of, 35; impact of, 35
group identity, collective will dominance of, 40
group meetings, interactive, in villages, 19

INDEX

Hashmi, Safdar, 101–03; untimely death of, 102
health, adolescent, 172, 181; and family welfare communication, 152; care, 270; care, sincerity of the, workers, 168; pregnant women's, 265; reproductive, services, 161, 162; women's, and nutrition, 190
Hindi-speaking states, 33, 104
Hindustan Latex Limited (HLL), 187
hoardings, 156, 180; at bus stops, 128
household industry, 40
human capital, 270
human resources, lack of, 270

International Conference on Population and Development (ICPD), 19, 158, 168, 194; recommendations of, 158
Information, Education and Communication (IEC), 149–50; activities, 171, 183; activities and inputs, 165; campaign, 186; decentralised, action plan, 183; effectiveness of, 182; implementation plan, 188; initiatives, 191; Orissa model of, 197; programmes, 171; state-specific, strategies, 172; strategy, 175; strategy development for family welfare, 150; strategy recommendations, 152; training, 192
illiteracy, eradication of adult, 81; eradication of poverty and, 69
illiterates, movement for, 117
immunisation, childhood, 66; promotion of childhood, 162
Improved Pace and Content of Learning (IPCL), 103, 114
India 2020, true vision of, 274
Indian Constitution, 267
Indian Context of Communication for Development, 29
Indian Forests Act, 244
Indian Institute of Mass Communication (IIMC), 192

Indian National Satellite System II (INSAT II), 205
Indian People's Theatre Association (IPTA), 137
Institute of Social Studies Trust (ISST), 203
Indian Space Research Organisation (ISRO), 52, 201–02
Indira Awas Yojana (IAY), 219
Indira Gandhi Garibi Hatao Yojana or the District Poverty Initiative Programme (DPIP), 20, 231, 235
Indira Gandhi National Open University (IGNOU), 201
Indo-China war of 1962, 33
information, dissemination, 238; dissemination campaign, 34; people's right to, 264; technology services sector, 272; model, 28
injustice, social conscience against, and oppression, 138
INSAT satellite, 201, 205
Institute of Development Studies (IDS), 203
Intensive Sanitation Programme (ISP), 224–25, 229
interlocutor-media-interlocutor, 46
interpersonal communication (IPC), 19, 38, 44, 86, 148, 153, 156–57, 169; skills, 171; skills of frontline workers, 175
Intra Uterine Device (IUD), 167
investments, public sector, in industrialisation, 30

Jaag Sakhi, 91
jamabandi transactions, 216
Jan Sunwaii (public hearing) programme, 264
Jana Natya Manch, 101–02, 137
Janadhikara kalajatha, 208, 216
Janata Party government, 65
jatha, 132; cultural, 83; proclamation, 128; troupes, 129; women's, 83
Jharkhand Forest and Environment Department (JFED), 20, 246, 248–49; bona fide intent of, 250

Jharkhand, 20, 245
Joint Forest Management (JFM), 20, 245–47, 249; coverage of, 146; implementation of, 248
justice, administration of, 274

Kadutham (letter), 131
Kahani Nahanachi, 61–62
Kalajatha, and environment building, 138; and folk theatre, 216; literacy, 129; local and traditional theatre performances, 19, 63, 122, 126, 131–34, 216, 241; performance, 129–30, 138; performance in the villages, 128; social mobilisation in terms of, 139; success of, 129, 138; success of KSSP, effort in Kerala, 63
Kerala experience of decentralised planning, 213
Kerala Shastra Sahitya Parishad (KSSP), 63, 82, 129
Kheda Rural Television, project, 52–55, 57; production unit for, 53
Kheda team: from SAC, 53; scriptwriter of, 53
Khilti Kaliyan, 91–92, 95–101; character in, 97; impact of, 100; women in, 98
Krishi Vigyan Kendras, 34

Land Reforms Act, 218
language, and culture, 43; non-official, 33; regional, 33; spoken, 33
latrines, construction of, 269; low-cost, 74; marketing of individual sanitary, in West Bengal, 257
leadership, administrative and political, 261
learner, adult, 87; adult, (poor rural women), 90; Adult Education Centre (AEC), 104, 106–09
learning, adult, 96
legislations, progressive, regarding the rights of women, 268
literacy, 26, 91, 117; adult, 35, 84, 86, 91, 114, 123; adult, and population, 19; adult, functionaries, 122; adult, programme, 45, 117, 126; campaign, 82–83, 133, 136; elderly woman attending the, class, 98; festival *jatha*, 128; functional, 35; instructors, 96; night, classes, 120; primer, 87, 89, 96; primer for women, 100; programme, 46, 90; realistic literacy curriculum for, 90; slogans, 128, 130; teaching, 129; use of, slogans in official communications, 128; village-level committees, 128
livelihood, sustainable economic, 269
low power transmitters, 31, 69

Mahila Dakiya (later called *Khabar Lahariyan*), 142
Mahila Mela, 60
Mahila Samakhya, 203
Mahila Samiti, 94, 97; formation of, 96
Mahila Swasthya Sangh (MSS), 187; members, 167; scheme, 164
Mahiti Astra, 213
marketing, social, 74
marriage, adolescent, 196; practice of child, 268
mass communication, modern, systems, 26
Mass Education and Media (MEM), 35, 147, 198
mass media, 19, 25, 38, 83, 86, 156–57, 165, 186; and literacy, 35; campaign, 120; communication campaign, 258; exposure to, 160; extensive use of, 27; funds used for, 186; materials, 43; misuse of a centralised, system, 37; reliance on, 148; role of, in national development, 30; role of, in the modernisation process, 25; role of, systems, 26; strength of, 28; use, 26, 44, 148
Mass Programme of Functional Literacy (MPFL), 125
Maternal and Child Health (MCH) issues, 153
Maternal Mortality Rate (MMR), 187

INDEX

media, access to, 17; adaptability of the folk, 136; audiovisual, 56; audiovisual, production techniques, 56; distribution of audiovisual, materials, 36; electronic, 111; government-owned, 35; local, 83, 180; materials (posters and charts), 182; national, campaign, 161; print, 258; reporting, 264; traditional, 43–44; use of audiovisual, 49
medical practitioner, local private, 261
Ministry of Human Resource Development (MHRD), 183–84
Minimum Needs Programme (Eleventh Finance Commission), 224
Ministry of Environment and Forests (MOEF), 245
Ministry of Health and Family Welfare (MOHFW), 150–55, 158, 165, 168, 172–73, 180, 182–85, 187, 189, 192–193
Ministry of Information and Broadcasting, 34, 165, 193
mobilisation: social, 66–67, 73, 83, 127, 136, 183, 185; social, and Information, Education and Communication (IEC), 170; social, effort through Zila Saaksharta Samiti (ZSS), 198
modernisation, 33, 41; international thinking on, 30
mortality, maternal, 196, 270; neonatal, 270
movement, mass people's, 116; women's, 59, 65, 70, 89; women's, in India, 63
Murzai Department, 218

Nai Pahal, 103; broadcast, 107
Namma Panchayati Namma Nirdhare, 213
Narmada Bachao Andolan (NBA), 64, 269
National Cadet Corps (NCC), 117
National Council of Education Research and Training (NCERT), 104, 107

National Elementary Education Mission (NEEM), 123
National Family Health Survey, 19, 164, 166, 173, 196
National Family Welfare Programme (NFWP), 151
National Film Development Corporation (NFDC), 193
National Film Institute (later renamed Film and Television Institute of India), 56
National Forest Policy, 244
National Health Policy, 194
National Immunisation Days, 73
National Institute of Adult Education (NIAE), 136
National Institute of Applied Human Research and Development (NIAHRD), 188
National Institute of Design (NID), 190
National Institute of Health and Family Welfare (NIHFW), 188, 190
National Literacy Mission (NLM), 19, 67, 81, 84, 101–04, 114, 116, 121, 133, 138, 184; document, 84; norm of bilingual primers, 104; programme strategy, 125; symbol, 128
National Literacy Mission Authority (NLMA), 130
National Population Policy (NPP), 150, 194
National Social Service (NSS), 117
Naxalbari uprising, 44
neoliterates, 140–41; weekly broadsheet for, 141
Networks and Movements, 59
New Policy on Education (NPE), 81, 88, 114
newspapers, 25; advertisements and radio and TV spots, 260; penetration of, into rural areas, 140; regular, 140; rural, 140
Nirodh condoms, commercial distribution and sale of, 147; marketing of, 147
non-literates, adult, 136

non-party political forces (NPPF), 66
Nyerere, Julius, 50

objective, of Project in Radio Education for Adult Literacy (PREAL), 104; of District Poverty Initiative Programme (DPIP), 232; of National Literacy Mission (NLM), 84, 86, 91; of the IEC component in the RCH Programme, 175; renewed, of NLM, 87
Ogilvy & Mather, 115
Oleum gas leak disaster, 63
one-way-video-two-way-audio, 204
Operation Blackboard (OB), 81

Padhna Likhna Seekho (learn to read and write), 83, 101–02
Panchayat Samiti, 266
Panchayati Raj, history of, 202, 212; Institutions (PRIs), 18, 150
Panchayati Raj (Extension of Scheduled Areas) Act of 1996, 246
paradigm, dominant, 27, 41–42; new, 158; shift, 175
parenting, reproductive behaviour and, 161
Participatory Forest Management (PFM), 251
Partition, 29, 33
pesticides, 34
Public Health Engineering Department (PHED), 224
Pitroda, Sam, 66
planning, and development, 41; of IEC, 182
play, and songs developed for the *kalajathas*, 217; on Mrityubhoj, 137; street-corner, 60
PNDT Act 1994, 195
Population Research Centres (PRCs), 184
population, growth of, 34, 195; India's, programme, 19
portapak, 53
Post Box 9999, 119–21
poverty, alleviation programme, 20; and deprivation, 34; and illiteracy, 38; rural, 44; social discrimination and, 35
P-Process, 76
prasav griha (clean delivery room), 240
Pratham, 267; experience in primary education, 266
Pregnancies, prevention of unwanted, 162; problems of, and childbirth, 62; reduction of, 270
Press, briefings, 182; censorship, 37, 65
Primary Education, universalisation of, 66
primary health care, 262; services, 171
Primary Health Centre (PHC), 259
printing press, advent of, 25
programme support communication (PSC), 73
progress, trajectory of, in the dominant paradigm, 27
Project in Radio Education for Adult Literacy (PREAL), 103, 105, 107–08; effectiveness of, 109; exposure to, broadcasts, 109; management of, 109; repeat broadcast of the, lessons, 111
Public Health Engineering Department (PHED), 222
publicity, campaign, 260; field, events, 182; field, programme, 182
Pudukkottai Case Study, 130
Pulse Polio Immunisation (PPI), 73

radio, 25, 29, 31; and film, 36; and television, 120; and TV stations, 43; broadcast, 55, 217; broadcasting, 29; in India, 55; lessons, 106; listening, dispersion of, 42; presenter, 107; programmes, 104; reader, supplementary, 104; receiver sets, 29; transmitters, 34; use of, 32
Radio Rural Forums, 31–32; success of the, experiment in Poona, 32
radio station, 29; broadcasting radio rural forums, 31–32; programme organisers in, 31
Rajasthan IEC Bureau model, 197

INDEX

Rajiv Gandhi Drinking Water Schemes, 218
Ramakrishna Mission Lok Shiksha Parishad (RKMLSP), 230
Rangayana, 203
Recognition of Forest Rights, 275
Red Triangle symbol for family planning, 36, 147
Report of the Working Group on the Status of Women, 89
Reproductive and Child Health (RCH), 19, 158, 166, 193, 197; adoption of the, approach, 166; awareness regarding, 185; communication challenge for behaviour change, 173; Communication Strategy, 192; components of the New, Communication Framework, 170; components of, 173; delivery of, services, 171; Household Survey, 173, 186; information decisions concerning, 169; National Communication Strategy for, 168–69, 172–73, 186, 189, 197; National, Communication Strategy Framework, 175; paradigm, 168; programme, 165, 167, 168, 173, 183, 185, 197, 199; programme management, 190; project, 162; Special Scheme for IEC, 187; state-specific Communication, Strategy, 171
Reproductive Tract Infections/Sexually Transmitted Diseases (RTIs/STDs), 162; treatment, 181
rights, people's, over the forest, 244; traditional, of access to the forests, 20; women's, 66
Right to Information Bill, 275
Rural Sanitary Mart (RSM), 223, 228, 229

South Asian Association for Regional Cooperation (SAARC) Year for the Girl Child, 82
Safdar Hashmi Memorial Trust (SAHMAT), 102
safe motherhood (prevention of maternal and infant mortality), 162
Samudaya, 137, 203
Sanitation Cell, 223
sanitation: reforms, 222; rural, 20
Sarabhai, Vikram, 37
Saraswati, 132
Satellite Communication (SATCOM), 208; facility, 203–04; format of TV viewing, 204; impact of the, 217; interactive TV, 204; lessons from, for rural application, 204; potential of the, system, 204; studio at ANSSIRD in Mysore, 220; success of the pilot, training experiment, 205; training, 204; training programme, 203
Satellite Instructional Television Experiment (SITE), 17, 36–38, 56, 201
satellite, installation of TV monitors and, receiver terminals, 205; receiver terminals, 202; transmission via, 205
Scheduled Castes (SCs), 54, 85, 217
Scheduled Tribes (STs), 85, 217
Scheduled Tribes (Recognition of Forest Rights) Bill 2005, 275
School Sanitation Programme, 224, 226
SEARCH, 203
self-help groups (SHGs), 189
service delivery, marketing approach to, 257
service providers, 260, 262
sex-determination tests, 172
Shiksha Karmi scheme in Rajasthan, 265
Silent Valley movement, 63
'sisterhood', 98
Satellite Instructional Television Experiment (SITE), limitations of, 52; project, 111
small family norm, 36
social animator, 45, 46
social mobility, ideas of, 26
social mobilisation, approach towards, 67; through the BGVJ, 139
Social Work and Research Centre (SWRC), 60

socialism, Soviet-type, 30
Song and Drama Division, 43
Sony Corporation, 56
Source–Message–Channel–Receiver (SMCR) model, 42; transmission/persuasion, 44
Space Application Centre (SAC), 52
State Family Welfare Departments, 184
State Forest Department, 266
State Health and Family Welfare Departments, 184
State Institute for Panchayats and Rural Development (SIPRD), 223, 230
State Institute of Health and Family Welfare (SIHFW), 188, 197
State Resource Centres (SRCs), 184
State Sanitation Cell, 226, 227
sterilisation, 149; enforced, campaign in the 1970s, 270; mass, camps, 37, 148
street theatre, 137
Swachagrama, 218
Swarnajayanti Gram Swarozgar Yojana (SGSY), 217

Tanzania Year 16 (TY 16) project, 50
Technology Information and Forecasting and Assessment Council (TIFAC), 272
Technology Mission, 66, 81, 84, 108, 114–15
technology, advancement in, application, 201
theatre, traditional, 42
THINK campaign, 147
Third World countries, 40
Total Literacy Campaign (TLC), 83, 102, 108, 121–23, 183; evolution of, 19, 125–26; expansion of the, programme, 139; in Kerala, 126; success of the, 139; success stories from the, to the press, 128
Total Sanitation Campaign (TSC), 221, 224
traditionalism, 26; transition from, to modernisation, 27
trained birth attendants (TBAs), 171
training, and capacity building, 237; in communication skills, 87; in IPC skills, 182; institutions, 191; interactive, programme, 210; of field personnel in communication skills, 72; programme initiated by ANSSIRD, 206; satellite-based, system, 20
transistor radio sets, advent of, 42
tribal, bilingual reader, 107; communities, needs of, 20; right to self-rule by, 245; trust between the, and the JFED, 249
TV, and radio, 156; channel, community, 55; channels, local, 156; serial, 92, 95, 205; signals for receptions, 205; spot, 193; studio, 205; transmission, 217

UN World Population Conference, 59
UN Conference on the Human Environment, 59
underdevelopment, in the poor countries, 40; of Third World nations, 40
unemployment, educated, 44
Union Carbide, plant in Bhopal, 63; accident in, 63
United Nations Educational Scientific and Cultural Organisation (UNESCO), 31, 43, 57
United Nations Children's Fund (UNICEF), 44, 57, 71, 91, 107, 115, 123, 202, 223–26, 228–29
United Nations Population Fund (UNFPA), 158
United States Agency for International Development (USAID), 151
Universal Programme of Immunisation (UPI), 71
Universalisation of Elementary Education (UEE), 66, 84
universalising primary education, 114
University Adult and Continuing Education Departments, 115
urbanisation, 26; industrial development and, 25

Velugu Bata, 141–42
video, 50, 57; audiovisual and, programmes, 87; camera, black-and-white, 56; courses for the use of video equipment, 51; for Farmer's Training in Peru, 51; portable, equipment, 50, 53–56, production unit, 51; recorder, portable half-inch, 56; use of, as a participatory tool for development, 49
vigilance groups, neighbourhood, 60
village audiences, 31
village development scheme, 271
Village Eco-Development Committee (VEDC), 246
village education committee (VEC), 124, 224, 265
Village Forest Management and Protection Committee (VFMPC), 246
village, motivator, 223
voluntary agencies in Rajasthan, 60

wall writing, 128, 185; reach of, and hoardings, 156
West Bengal Rural Sanitation Programme, 222, 225, 260
women, activists, middle-class, 60; and children health issues, 173; as learners, 88, 91; in distress, 60; in the films, 92; learners, 92; married, 165; meeting with, 238; portraiture of, 92; portrayal of, in cinema, 60; portrayal of, in the media, 62; rural, 60; status of, 89; stereotypical portrayal of, 92; television programmes for rural, 91
Women's Liberation Movement, 62
workers, frontline, 165, 183
World Bank, 193
World Health Organisation (WHO), 44, 57
World Social Summit, 272

Xavier Institute of Communication Arts in Bombay, 56

Zila Saksharta Samitis (ZSSs) or District Literacy Committees, 139, 180, 183–85; evaluation of the, experience, 184; NGO partners of, 185; scheme, 185–86

About the Author

Avik Ghosh is an independent consultant and currently works, among others, for the Observer Research Foundation, UNICEF and the World Bank. Having obtained a master's degree in economics from the University of Delhi (1969), he began his professional career as a producer in All India Radio. He was one of the founders of the Centre for Development of Instructional Technology (CENDIT) in 1972, a pioneer non-governmental institution promoting the application of audiovisual media and video technology in education, training and rural development.

Avik Ghosh subsequently joined the National Literacy Mission (NLM) and worked as a media consultant in the Directorate of Adult Education, New Delhi (1988–91). He was then a Senior Fellow at the National Institute of Adult Education (NIAE) from 1991 to 2003.

HN690.Z9 I564 2006

Ghosh, Avik.

Communication technology and human development : 2006.

2008 01 11

0 1341 1049084 1